T0003321

FORGET ME NOT

FORGET ME NOT

Finding the Forgotten Species of Climate-change Britain

Sophie Pavelle

B L O O M S B U R Y
LONDON • OXFORD • NEW YORK • NEW DELHI • SYDNEY

For my parents.

BLOOMSBURY WILDLIFE
Bloomsbury Publishing Plc
50 Bedford Square, London, WC1B 3DP, UK
29 Earlsfort Terrace, Dublin 2, Ireland

BLOOMSBURY, BLOOMSBURY WILDLIFE and the Diana logo are trademarks of Bloomsbury Publishing Plc

First published in the United Kingdom 2022

For legal purposes the Acknowledgements on p. 337 constitute an extension of this copyright page

A catalogue record for this book is available from the British Library.
Library of Congress Cataloguing-in-Publication data has been applied for.

ISBN: Hardback: 978-1-4729-8621-4; Audio download: 978-1-3994-0019-0;
ePub: 978-1-4729-8622-1; ePDF: 978-1-3994-0620-8

2 4 6 8 10 9 7 5 3 1

Typeset by Deanta Global Publishing Services, Chennai, India
Printed and bound in Great Britain by CPI Group (UK) Ltd, Croydon CR0 4YY

Contents

Prologue

Shall we?

Nature has always been there. A lovely little addition to our lives when we want it. A sunset? Yes, please! A boat trip to see some whales? Count me in! A kingfisher flying past during a picnic? Hello! But nature is challenging us. Actually, planet Earth's climate and biodiversity crisis removed the cushion from underneath us decades ago. Yet the strange thing is, we still don't seem to have noticed.

As I write this in the summer of 2021, the past two weeks have been fraught with change. Germany, Belgium and Uganda are dealing with horrifying, deadly floods. North America is grappling with record-breaking temperatures for a second time this year. Siberia – reliably one of the coldest places on Earth – is battling unprecedented wildfires. Back here in the UK, the Met Office has issued its first-ever extreme heat warning, and the government is proposing a new oil field in the North Sea. All this while one in every seven UK species is facing extinction. And global ocean plastic is set to triple by 2040, with plastic items outnumbering fish by 2050. How far are we willing to push our planet before it's too late to turn back?

We live in an era of contradiction. We're being told where we can and cannot go, what we can and cannot do to the environment. Red tape and paperwork intended to save nature are, in fact, pushing us away from it – and away from each other. Doesn't it sound exhausting? I don't know about you, but I find it so hard to know what to think, who to trust or what to do about it that I switch off from it all. Humans are striving to have the last word, but we're becoming lost in the process.

I have written this book because I'm worried that we'll forget what we're losing. I'm worried that we're moving too fast within these lives we've built for ourselves, at the expense of what makes it worth living on this planet. The species that give life meaning. The species that got here first. The word 'forget' has origins in ancient Germanic prose and loosely translates as the act of 'losing grip' or, more commonly, 'to lose care for'. This is what is happening around the world, around the UK, around where you live: nature is waving red flags at us, sounding alarms and blaring sirens to try and get us to listen. As I see it, we have two options: we can stand by, ~~explore our bodyweight in wine~~ and continue to enjoy a rather depressing show. Or we can sit up, trust in ourselves, work together, and (quickly) try and do something about it.

In 2019, María Fernanda Espinosa Garcés, President of the 73rd United Nations General Assembly, opened her keynote with the announcement that humans had just 11 years left if we want to avert a climate catastrophe. But surely it will be alright? Don't things always work out? Like, 'If I have oat milk in my reusable coffee cup, then I'm saving the planet … aren't I?'

The British government has ambitions that the UK will transform into a carbon-neutral economy by 2050, playing our part to ensure that global temperatures don't rise more than 1.5°C. Our signature on the 2016 Paris Climate Agreement legally requires us to meet these obligations. In 2021, the UK hosted the G7 Summit for global leaders and the 26th United Nations Climate Change

Conference (COP26). We might be making ourselves feel better, but I'm unconvinced that the impact of climate change is being taken seriously. We still prefer playing fast and loose.

As with all crises, we should be jarred by this. But instead, of course, we distract ourselves. The economy, politics, global health, food security, opinion polls, human rights, equality, social media, football … we're intelligent mammals trying to evolve within our world of veneers. We know it isn't sustainable. It never was. Yet we're fantastically missing the point: nature *is* our economy. And that means nature is, also, unfortunately, political. The climate crisis is a human rights crisis, an equality crisis. Climate justice is social justice. Nature's health ensures our health. We're kidding ourselves that we are chairing this meeting when in reality, we are nothing more than participants (or senior board members, at a stretch).

It sounds scary because it is. I could reel off more dispiriting data, but I know that's not why you're here. I could ask you to panic, but that won't help either. Good work is rarely accomplished in a panicked state of mind. Decisions are likely to be poor and hastily made. We won't be economically brave. So, no, I don't want you to panic. I just want you to *know* nature and to urgently prioritise that.

To help you with that, I would like to introduce you to a few species, habitats and landscapes that we take for granted, some that you might never have heard of. Because bothering to change the way we live will *only* work if we are invested in the outcome. I believe that real action, progress and hope begins with a celebration and a better understanding of our place in the game. If we want to continue to be residents, we need to take a step back and better understand our *home*. Call me naïve, fanciful, unrealistic – I honestly don't care – but I believe it's only when we work on restoring ourselves that nature will be able to do the same. It's all or nothing.

I've always enjoyed exploring and having adventures – especially across the UK. I have my parents to thank for that, and I'm aware of what an enormous privilege it is for the outdoors to be a huge part of my life. But nature writing and travel books often frustrate me. I'm not a voracious reader and find it hard to concentrate on texts for long periods. Sometimes it's easier to read about things that don't exist. But I feel put out when I read some books which present a glorified – and dare I say, typically *male* – quest through the natural world. A world in which everything is wondrous, and male naturalist travelled to see 1 million plants/birds/butterflies/moths (and of course, succeeded).

Other times, I've felt I can only *truly* connect with nature if I've suffered a traumatic experience. That nature can only serve me if I hit rock bottom. Of course, science shows us that nature can shed light into our darkest hours. I can vouch for this. But what about experiencing nature for the sheer hell of it? The irreverent joy of trying to find wildlife – and failing? The simplicity of totally winging it and being more in the moment?

I am not a naturalist. I can't tell you I have a deeply profound connection to nature; I just like it, and I enjoy learning and working alongside people who know a lot more about it than I do. No, I don't know how to watch birds properly. And yes, at times, I would rather watch *Love Island* than a nature documentary. I would choose pub over wildlife hide. Social media has made me demanding, anxious and shortened my attention span to dangerously low levels. I am infamously terrible at knowing what plant or animal I'm looking at and would happily lump all gull species together and refer to them as 'seagulls'. I have a degree in an envelope somewhere, and yet I still cannot sermonise the delicate nuances between chiffchaff and willow warbler. But that doesn't bother me. I don't think it's important. For me, this 'nature stuff' is more about the journey. Nature doesn't care where you come from or what you look like. The journey doesn't have to mean anything at all. Having the audacity to engage with it in your own way is absolutely enough.

During my year of writing and travelling, a lot was going on in the environmental sector in the UK. It was stressful, if impossible at times, to keep up with and understand. Yet, the reality of it all only bolstered my hopes for this story. The UK is like nowhere else. For a start, the British public muddle through challenges with wry, disarming humour. We prescribe a cup of tea as a universal antidote. We trigger family rifts when debating whether jam or cream should go first on a scone. We are a nation of (mostly) good people who apologise a lot and consistently achieve more when we come together. We have a tapestry of landscapes and natural history of which we should be unbelievably proud, and to which we should offer more of our attention.

Let's not forget that the UK is an archipelago. I still feel excited every time I realise that. You can be on a mountaintop in the morning and riding the Circle line around London by evening.

Capitalising on this geography, I wanted to experience first-hand how ready the UK is to transition into this 'carbon-neutral economy'. Are the wildest corners of the British Isles feasibly accessible by greener modes of transport? Is low-carbon travel even realistic? Aside from hoping to maintain a cracking tan, I spent a year making these 10 trips via bike, boot, a ridiculous number of trains, ferries, a kayak and an electric car in a (somewhat haphazard) attempt to see how sustainable travel can be from the depths of Cornwall to the heights of Scotland. I discovered the challenges of remaining loyal to travelling sustainably, and wanted to know whether it was something we needed to start taking much more seriously. Besides, I couldn't very well write a book about nature's resilience to human-induced climate change from the driver's seat of my car.

A story is more enjoyable with characters, so I've chosen 10 stars to lead the narrative. An impossible task, as there are many more than 10 animals and habitats in need of our serious attention! But I have chosen these for their modesty. They are not your average poster children, and I confess even I barely knew anything about them before I started probing. I also chose them because I knew they would allow me to highlight a decent portion of the complex environmental issues threatening the UK and the world. And, side note, I've fallen hopelessly in love with them.

Like the grey long-eared bat, some species hover on the brink of extinction and climate change may be their final push off the precipice. Others, like dung beetles, may fare better in a warming world than we thought. And the rest? Well, they're just downright awkward, and solving sentences describing their predicaments has kept me awake at night. (Merlin and harbour porpoise, I am looking at you.)

But all 10 characters are messengers that we would be unwise to ignore. All 10 have more of a right to exist in the British Isles than we ever will. It's a privilege to be alive at the same time as them. This book is my tribute to these species and their habitats. It's also my tribute to science and the utterly brilliant, brave people fighting this essential fight. And it's the honour of my life to have the opportunity to tell this story.

I wrote this book during the world's most significant crisis for a generation. Yes, Covid-19 presented some, um, *interesting* hurdles to overcome, but I never planned on this being a 'pandemic book'. And between you and me, I hope it goes beyond that. However, what the pandemic has done (more effectively than any campaign, film or petition) is reveal truths that, until then, had been massively economised, overlooked, and perhaps even forgotten. Being forced to fight for our survival has been an overdue awakening: that nature is part of us and its survival is our own.

It occurs to me that crises like climate change and biodiversity loss should bring out the best in humanity. We only have to look to our past and present to see that we *can* do this. The Blitz, terrorist attacks and the pandemic have triggered our innate instinct to do good and realise the immense capability of the human species. And that all has a better chance of happening if we are *emotionally* aware of the planet and what shares it with us. We grieve harder and longer for family members than we ever could for total strangers. It's only human, right?

Some people exist in this world as mediators, desperate fixers, go-betweens. I reckon I am one of those people. But we live in an era that also requires those people to be disruptors, because it's all hands on deck in this extraordinary world. And I don't know about you, but I find this prospect quite exciting. I hope I can be all of these things to you over the 10 chapters that follow. Thank you for being here. Together, we'll journey to find those not to be forgotten. We all need to be the verb. This book is my attempt.

CHAPTER ONE

Marsh Fritillary

It wasn't that long ago when seeing a butterfly was enough. There it is – that was nice! Then crack on. But now? It's all about advancing the public archive. Smartphones materialise in some performative act of connecting to 'The Nature'. Fine. And yet, I can't help but wonder whether it's because the natural world plays such a minor role in modern life, and encountering it is suddenly a novelty that must be immediately documented and preserved. Thou shalt seek the path with the best view and record *crucial evidence* of doing so. I was cycling through a park near home in Exeter when a runner made a concerted effort to stop, flamboyantly tapping a watch on her wrist. She quickly photographed a swan as it preened beside a patch of graffiti, which asked onlookers, 'wHAt DoES yOUR *SoUL LOok liKe?*' I smiled. I would have done the same.

Many people in the park were exercising with great determination. Faces wore hardboiled but jubilant expressions, as though the rush of endorphins and Lycra-assisted lunges were collectively giving Covid-19 the finger. I was on my bike and nodded to some other cyclists I passed on the path, the token fist bump of, 'I see you (but only if I like your bike).' It was June 2020, and the first lockdown due to the pandemic still blanketed the UK. But thankfully, the restrictions in place still enabled me to head out on my bike to try and see a butterfly – a childish-dream-turned-liberating-novelty, when so many people remained cocooned within an anxious chrysalis.

I was late, spinning past blocks of university accommodation and enjoying being out on my bike again. Like most people in their mid-20s, I lead a chaotic existence. A bit anxious. Pretty selfish. Back home after studying, daunted by debt, deciding what to do next. Blah, blah. (Quite pretentious, too.) That day, however, I was *investigating*. Nervous at fleeing the sanitised safety of home, my survey started small – cycling to Exeter St Davids station to board a train to Bodmin, Cornwall. Woodpigeons played maverick in the warm sunshine. As I sped along, shards of sunlight escaped through full, verdant branches in those stressful rhythmic strobes that pervade the kind of nightclubs I went to once or twice. Peering over Exeter's Millennium Bridge, I spied another swan, a mother, serenely coiled on a raft nest. The father (as always with swans) was floating nearby, alongside a Lucozade bottle.

I arrived at the station in my latest incarnation as Hopeful Butterfly Whisperer. As I fumbled with hand sanitiser and tickets, the tannoy woman repeatedly thanked everyone for 'maintaining a safe distance from other passengers at all times' and 'keeping travel to an *absolute* minimum'.

This passive-aggressive '*absolute*' was personally directed at me. I was sure of it. Sheepish, I crouched, head hidden in my rucksack, conscience itching. Was going to see a butterfly via this 'low-carbon route' *really* necessary? My bike fell onto my shoulder like a provoking kick from a sibling under the table. The sun blazed overhead. A magpie spluttered into gear by the taxi rank, its plumage wickedly metallic as it stabbed a crisp wrapper with its bill. A robin assumed position atop the pillar box in a very 'shut-up-I-am-very-busy-and-important!' manner. Actually, I needed this. It *was* necessary, thanks.

Pulling away from the station, we began along my favourite stretch of track on the South Devon Main Line. After moving from America in 1998, our home for the following two years (a tiny riverside apartment) overlooked this huge stretch of the Exe Estuary. I still cannot believe our luck in kicking off childhood like this. I used to watch the caterpillar of train carriages from the lounge window as they traced the edge of my little world, like a giant game of Snake. I was four years old. Where was it going? Who was travelling? Why? Sometimes, if the tide was out, the carriage lights would reflect in the mudflats.

Twenty-one years on, and I was a passenger on that train, winding west. Individual trees, dips and rises of the land were as familiar to me as my own freckles. The tide was in, and a grey heron paused on the edge of a flooded pool as though testing the water's temperature. Beyond Dawlish especially, window-seated passengers have the illusion of flying over the sea, so close do the tracks run along the sea wall. I'm pleased to admit that I've got a lot of time for the River Exe and its bottomless pit of nostalgia. I will never have enough of it.

Safe to say that after weeks of lockdown, the train was particularly thrilling. It felt criminal to be out. I had escaped and was on the run. Tannoy woman was no doubt raging on the platform back in Exeter, but there was nothing she could do now.

My favourite mode of transport, a striking white, red and black gravel bike, was wedged loyally next to me in the bike rack. I kept stealing glances at it like we fancied each other. We were set to have quite the year together. Through the window near Teignmouth, the sea looked soft and lumpy like an unmade bed. I wanted to sit in it. An idle column of iconic red sandstone braced the swell. Picking up more escapees in Totnes and Ivybridge, rivers turned to fields, flirted with moors and returned to fields. We sped through tunnels, and masked reflections peered back at us from black windows – our new selves. I studied my face as though it belonged to someone else.

Before I left home, a lady called Jo Poland from the Cornwall branch of Butterfly Conservation told me about an area of Bodmin Moor favoured by one of the UK's rarest insects – the marsh fritillary butterfly, also known as *Euphydryas aurinia*. As you'll find out, the marsh fritillary is a gorgeous little thing. But it doesn't exactly help itself. Its penchant for moist and tussocky landscapes has largely confined it to western areas of the UK in erratic numbers, where this habitat is precariously holding on following a problematic agricultural past (hold that thought ...).

Thriving in groups of highly connected colonies that radiate across highly connected habitats, the marsh fritillary generally sticks to flight distances of not much more than 50–100 metres – and lives the definition of a 'social butterfly'. Hard and fast promiscuity within this insect's environment is the nexus of its survival. Good quality connected habitat is the name of the game – the Union to the Jack. That said, it's a butterfly that is both nowhere and everywhere. Although local populations remain so, as a species, it has curated a lifestyle to suit the climes of 38 countries in Western

Europe, North Africa, Asia and Korea. The marsh fritillary strikes me as a cultured cosmopolitan.

Nearing Plymouth, queues of cars waited for green lights, like cows ready to be milked. Houses. More houses. Grey upon grey. From our vantage point crossing the Tamar Bridge into Cornwall, I watched as small boats and cars moved slowly towards the estuary mouth, like accessories to a giant train set. The thing about Bodmin Moor is that it's one of those places that is very easy to overlook. Most people on their beeline to Cornwall simply whizz through Bodmin with a wee-stop and a picnic. (Me included.) But Bodmin deserves a lot more than that. A colourful history decorates this bleak, windswept land with Neolithic and Bronze Age hut circles, ruins of medieval chapels and the fabled resting place of Excalibur at the bottom of Dozmary Pool. Granite outcrops and stone circles sew a rich heath upland into an Area of Outstanding Natural Beauty, which is a bit of waffle for a habitat nearing the Top of the Aesthetic Class. Daphne du Maurier's *Jamaica Inn* thriller was born here, as was the pub of the same name. The River Fowey rises high on the moor, just north-west of Cornwall's tallest hill – Brown Willy, meaning, 'Hill of Swallows' (sure). Pulling into Bodmin Parkway station, I realised I had never been here before.

I set off along National Cycle Route 3 – a portion of the 328-mile cycle safari from Land's End to Bristol, through Cornwall, Devon and Somerset. The National Cycle Network is identified by its blue signs, shining beacons of encouragement (or despair) for long-distance riders. In 1984, the Avon-hugging 15 miles linking Bristol and Bath on the disused railway was crowned as the first 'route' in the National Cycle Network. Now in the custody of the UK's walking and cycling charity Sustrans, the network has grown to cover ten main national routes, ranging from the Shetland Isles to Dublin and St Austell, across 5,273 miles of engineless trails and 11,302 miles on some sort of road. Thankfully, my butterfly ride was just a wholesome 10 miles, and no fewer than three separate signs directed me and my bike to the trail.

Swooping under the rail bridge, I happened upon a dreamy scene straight out of Kenneth Grahame's[1] head. Petals had fallen from a vast magnolia. I flew down the pastel pink aisle, beckoned by the wedding of woodland arching over my helmet. When my Nanna died in 2002, the magnolias were in bloom. Now, every time I pass a magnolia with my mum, I know that she's thinking of her. I think of her too. Further into this *Wind in the Willows* scene, a young rabbit inspected his front paw, and behind him, a Cornish vista unfurled. A greenfinch sounded like someone was trying to whistle and hum at the same time. Two ladybirds embraced on a nearby nettle. Good times. I was high up on the path over a bridge, and the River Fowey rushed below, too busy to stop. Perhaps this is what England once was. Provincial, spacious, deeply wooded – but not obsessively so. An enormous pastoral meadow percolated with oak, ash and beech introduced the sweeping grounds of the National Trust's Lanhydrock Estate. The whole scene felt clichéd, strangely familiar. I liked it.

It is everything you might expect from a grand Victorian country estate. A lavish history of housing high-society mid-1600s Parliamentarians was enjoyed before a devastating fire in 1881. Later, its slate-and-granite walls sheltered evacuees and woodland ammunition stores during the Second World War (the surrounding deciduous canopy offered ideal protection from enemy eyes). And then, in 1953, the seventh Viscount Clifden – great, great uncle of Ollie Williams, *Love Island* contestant, winter 2020 – gave

[1] Author of *The Wind in the Willows* (1908) – if you haven't read it, stop reading this and do that first. And then please come back.

Lanhydrock House and 160 hectares of parkland to the National Trust. The plot thickens. As in early 2020, Ollie exclaimed to the press that he was the rightful heir to the estate before realising that once a property is nestled within the loving bosom of the National Trust and its members, 'inheritance' is no longer a thing. Another time, babes. Today, this Victorian estate is famed for its balance of being wealthy yet largely unpretentious. *Not exactly*, I thought, as I passed the seventeenth-century stone folly and gatehouse.

A raven clattered about in some branches overhead as though clearing up after a big meal. Heavy mouth-breathing seemed to help shift gears and pull me up a steep hill. In its Terry's Chocolate Orange coat, a meadow brown butterfly flew with me as I drew Z-shapes up the hill. I still wonder whether it, too, was tracking our tango. At the top, I took off my leggings. *Had* to. The air felt impossibly thick, and I found myself standing, panting and trouser-less on the side of a Treffry Lane. For no reason at all, I ran across this Treffry Lane wearing just my pants – and ran back again. Perhaps in a confused bid to create some airflow and cool off? Hurriedly, I shimmied into cycling shorts and pretended that everything was fine.

Wending left on a satisfying hypotenuse across the estate, everything around me began to look very overdressed. Foxgloves adorned hedges like purple candelabras. The air smelled of festivals. Of Sundays and childhood. A carnival of spring colour with *Free Entry!* for all. Darting out from the right, a swallow conveniently flew down the next section of Route 3. A rubber-stamp approval from our most regal of migrants felt like a sure charm.

Beyond the hedge, I caught glimpses of where the butterfly was meant to be. High up on a hill to my left, the nature reserve emerged in a random blend of landscapes. Lying on the northern end of a granite ridge, Helman Tor is sandwiched within a celebrated area of conservation and history. One could say that it has done very well for itself, for it has quite the portfolio. The habitat surrounding the tor is one of the 57 nature reserves managed by Cornwall Wildlife Trust. Age has been kind to this area of land. It is now a County Geology Site (CGS), a Scheduled Ancient Monument (should we save the date?), and it lies adjacent to the Red Moor Site of Special Scientific Interest (SSSI) and Breney Common Special Area of Conservation (SAC). 'Tin-streaming' from open-cast mines sculpted the lay of this land for many years, including the setting of many of the ponds and woodlands, the scars of which can be easily seen and felt against the background hum of the nearby A30.

There's more. Above a haze of foxgloves so numerous that the moor glowed fuchsia, buzzards cruised the thermals like a lazy river. I felt ushered towards a heap of granite that looked like a pile of laundry. Skidding to the side of the car park below Helman Tor, I addressed a serious state of hunger while sitting beside my bike under a tree, musing.

I considered that if I were on a similar nature escapade, say 30 years earlier, insects would have featured heavily. It's one of those visions that people of my parents' generation sometimes refer to: 'We had butterflies everywhere!' and '*All* over the windscreen', 'SPLAT! they'd go – every *bloody* time!' And such. Thirty years ago in a given field, grasshoppers would have leapt ahead of me on my (*ever so* hearty) stroll. Midges and mayflies would have prompted the carefree hand-over-face waft – also typical. Fruit flies and other small things would have wriggled their way inside my clothes. Hoverflies may have occasionally taken a break on my shoulder. And I would have been OK with all of this. Perhaps I then would round a corner to discover a blossoming bramble bush, glittering with feeding meadow browns, speckled woods, red admirals, tortoiseshells and (while we're in Utopia) all eight fritillaries. A riot of tiny life being lived.

Yet, I glimpsed into that old world as I sat there, failing to remember ever seeing so many butterflies and bees in one small space. Verges were humming. Not because of the A30, but rather an invertebrate full house. Or so it felt. A peacock butterfly sunbathed on a bare patch of ground. Red upperwings as rich as the season, its eyespots unblinking. Quivering in luminous flame, a brimstone butterfly fed on a pink ball of clover. Nature's cosplay can be very convincing. I felt officially seduced. It turns out that this 500-acre (202-hectare) site is a veritable orgy of wet heathland, dry heathland, acid grassland, ponds, ancient oak and willow woodland – and loads of insects. Overall, it seemed I was experiencing a place where nature has been granted tenancy. That's why it felt different.

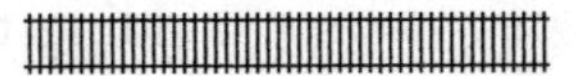

Yes, other endangered species are available, some in even greater peril. But there was something about the marsh fritillary that pulled me closer, maybe because it's considered a bit of a talisman across British moorlands. Or perhaps just because I liked the look of it. Either way, I was in.

Its UK distribution has declined by 75 per cent in 25 years (and the UK is still considered a stronghold). Sixty-six per cent of England's colonies were lost between 1990 and 2000, and now it exists in a fugitive state amid a devastating loss of habitat. Unsurprisingly, this tip-toeing across a fraying tightrope has turned this insect into an international symbol of resilience, unwittingly catapulting it into big-picture conservation like a person who achieves overnight fame simply by surviving a horrible experience. A little dark, I think. Wildly intrusive. But for some reason, with the marsh fritillary, I'm left feeling impressed. As though its world is slowly descending to ashes, and yet, still it rises like the phoenix.

Such a rough ride in the UK and across Europe has made the marsh fritillary a conservation priority, which has afforded it the very highest levels of protection. A UK Priority Species, it has been fully protected under the Wildlife and Countryside Act since 1998, and it remains under 'vulnerable' European status. Why not throw in two international statutes of protection while we're at it? All of which makes the marsh fritillary the most protected butterfly in the UK after the large copper and large blue.[2] It's safe to say that messing with this species will induce a legislative migraine.

It shouldn't come as a shock to learn that agriculture has been the biggest threat to all butterflies in the British Isles so far. In the south-west of England alone, the annual turnover of agriculture is more than £2.7 billion – more than any other region. Bidding farewell to 92 per cent of regional culm grassland (tussocky, moist, generally fabulous) by merging fields and eradicating natural buffers over the past 100 years has removed a vital climate change ally from South West geography. Culm grassland stores five times more water than monoculture grassland – as well as filtering it, reducing downstream flood risk and improving water quality. There's also the small bonus that the topsoil beneath the tussocks stores twice as much carbon as intensively managed farmland. And although 25 per cent of England's SSSIs are in the south-west, nearly all of these sensitive areas (including Dartmoor and Exmoor National Park) are farmed.

Once widespread throughout the British Isles, the marsh fritillary has been an unfortunate passenger on some of the most nauseating population rollercoasters of any species.

[2] FYI, the large copper is protected under European legislation but became extinct in Britain in 1851. The large blue became extinct in 1979 but has been successfully reintroduced to Somerset and Gloucestershire. Hooray.

Colonies have been lost from 79 per cent of locations marsh fritillaries once were, and since the 1980s and 90s, the species was known to breed only on a handful of sites. Its history is a textbook tale of sacrifice and tyranny – apt bedtime reading. It's also a story of what happens when we don't employ landscape-scale thinking in our merry bid to armour a world against elemental surprise (while ensuring we get to live in a world that can also deliver hot ramen to our doorsteps with a mere swipe and a click).

The 2019 *State of Nature* report, published by the National Biodiversity Network, cited butterflies and moths among the hardest hit of all species in the UK. Humanity booms, wildlife dwindles. Alas. This tits-up of a trade-off will crop up quite regularly as you read this book – but there's more to it. Since the start of the twentieth century, more than 77 per cent of UK land has been handed to agriculture and apparently improved (meaning a million hectares of wetland drainage, ploughing, reseeding, concreting rivers, having a *literal* field day with fertilisers … that sort of thing). And our bid to go all the way in our complicated relationship with carbon has removed nearly everything this butterfly needs from its ecosystem. So much so that in Devon and Cornwall, 8 per cent of the marsh fritillary's favourite habitat remains, squeezed into two of Britain's most popular tourist destinations. Both of these counties also happen to be gardened by a significant number of sheep, as you'll find out soon.

First, though, permission to objectify this butterfly. Picture for a second, if you will, a Victoria's Secret model on the runway – let's go with Gisele. Gisele is glamorous. As far as we know, naturally so. A proper specimen. Dripping with allure, etc. And that's what the marsh fritillary is – ahoy there! The undisputed cover star of the butterfly world. The name 'fritillary' is used for eight UK butterfly species, and stems from the Latin *fritillus* meaning 'chessboard', a title also bestowed upon the *Heliconiinae* butterfly family and *Fritillaria* flowers, the other chequered members of the natural world. Butterfly fritillaries are also collectively known as the 'checkerspots', which is just a delight and much easier to remember. Put it this way, these dancing, brightly coloured miniature stained-glass windows make for a substantially more aspirational rendezvous than our reliable meadow browns (bless their tiny hearts).

The best way to tell that what you think is a fritillary is indeed a fritillary is to look at the underwing. These act as a trademark for each species. The creamy, apricot-coloured underwing of our marsh fritillary embodies the genre of colada we might sip on a harbour wall in Majorca while comparing tans and anticipating the night ahead. Its upperwings contrast vividly – bold and strong, perfectly traced in a soft black – the perfect outfit to complement the cocktail. To put it mildly, you are looking at the most satisfying aesthetic of symmetry and charisma. Marsh fritillaries are generally brighter and more outrageous than their seven cousins – their iconic sunset-and-black tessellation is made even more 'haute couture' by a row of tiny black dots skirting the edges of their hindwing. Like our own transition from sweet 10-year-olds into don't-look-at-me-I-hate-everything teenagers, marsh fritillaries become shiny when their rainbow regalia fades. And in this state, they have been rather insultingly referred to as the 'greasy fritillary'.

As far as I knew, I had never seen one before – despite my childhood featuring damp, delicious Bodmin and Dartmoor, both of which are good areas for them. Sitting below Helman Tor for some time, grass imprinted the backs of my legs. I scoffed some leftover mange tout from dinner the night before with immense disappointment. A beautiful second sandwich and an array of other snacks were forgotten on the kitchen table back home. An eager tick, climbing the laces on my left shoe, fancied some warm, recently fed mammal – it was one of many that have chosen me over the years. It struck me that I could lie there and await a long, slow peel into fatigue as I satisfy the sanguine urges of this tiny arachnid. It also struck me that I ran the risk of becoming a pathetic narcissist, and I had better bloody go and find what I came for. So, as I made a beeline for the sweet spot Jo Poland had described and, when not checking every available crevice for ticks, my eyes were primed for flashes of orange.

This species was becoming hard to resist. Every person I have spoken to about this butterfly, be it scientist, volunteer or general butterfly devotee – has raved about its life cycle. From a young age, the lullaby of the butterfly life story is drummed into us (or jolly well should be). The late Eric Carle created *The Very Hungry Caterpillar* to show our pre-school selves (and adults) the thrill of transformation. Sometimes, it's the stories of the smallest beings that stay with us. Yet the marsh fritillary makes the typical egg-caterpillar-chrysalis-butterfly situation seem as dull as a Tuesday. Friends, this is a butterfly on steroids.

As Carle so beautifully wrote: all butterflies and moths have a four-stage, sequential life cycle: egg > caterpillar > chrysalis > winged adult. The marsh fritillary takes a year to waltz around that circle, and the caterpillar (larva) sheds its 'coat' five times – this process is 'moulting'. Each stage between moults is called an 'instar' – and the marsh fritillary has six. OK, so far? Good. Marsh fritillaries have one annual flight in May and June, and I was hoping to catch the tail-end of this period. Their development is chivvied along by sunshine and warmth in a similar logic to a ripening banana on a windowsill: the warmer the site, the faster they develop. As soon as the adult butterfly emerges from the chrysalis (male usually first), mating is on the agenda – sometimes within the hour. After months of solitary infertility, their newly winged selves tumble into a flirtatious world of interpretive dance and promiscuity. Everything nature (and university freshers?) wished for.

A marsh fritillary couple is arrestingly beautiful. We may envy their lottery of perfect genetics but surrender to their appeal like a hysterical fan desperate for an autograph. You may be surprised to know that the mating ritual of a marsh fritillary is a prudent, 'Don't look at meh!' affair. After a bit of impassioned zig-zagging, the male finds his equally eager lady. Pleasantries exchanged, the pair settle on a leaf perch and turn to face *away* from each other, joining only at the tips of their abdomens. These romantic rendezvous can, like any first date, last from minutes to hours. Next time you consider 'multitasking' as a problematic burden, it's worth noting that some butterflies can casually fend off passing birds mid-coitus.

Nosy beaks aside, the male deposits his sperm packet – a neat dollop with a dash of nectar-derived carbs and proteins for good measure. This little sexual picnic seals the deal and fertilises the eggs that lie inside our patient lady, permeating through tiny pores in the eggshell. Now, the challenge is for the female to transport her immense cargo of (in some cases) 500 fertilised eggs to a leaf of Devil's-bit scabious. Her abdomen is swollen and heavy, physically limiting how far she can fly. Eggs are always laid on the underside of the leaf, which generally needs to be large and cushioned to absorb the weight of hundreds of little life bundles.

For some people, this is a captivating moment. Dr Caroline Bulman, senior conservation manager for Butterfly Conservation and a leading authority on all things marsh and fritillary in the UK – is particularly taken with this part of its life cycle. 'The number of eggs is truly staggering,' Caroline smiled over Zoom. 'Wander along grassland in May and June, find some Devil's-bit scabious, turn the leaf, and you'll see this *gorgeous* pile of newly laid yellow eggs.'

Being in late June, my visit didn't yield any butter-yellow, perfectly spherical marsh fritillary eggs, which, when I looked them up, reminded me of miniaturised giant couscous. Each egg contains bundles of tiny cells, each cell primed to become wing scales, antennae or proboscis one day. With lives so entirely scripted, any butterfly we see demonstrates evolution at its most magnificent. At this point, the butter-yellow colour has darkened into a purply-brown, so the eggs now resemble giant couscous soaked in HP sauce. It takes around 30 days for their eggs to hatch, varying sometimes depending on whether early summer decides to be warm and sunny or wet and tragic. Once they hatch, the caterpillars emerge, and the silk webs are spun. Avengers, assemble.

I spent a happy hour wandering randomly all over the tor and enjoying the fact that it was Monday and I wasn't doing Monday things. The sun was hot, yet the sky seemed to be having anger issues. Rumours of a thunderstorm murmured across the land. I am no botanist, but the plant life here was extraordinary. A real botanist would swiftly become overexcited, if you know what I mean – sundews, sphagnum, pillwort, Western bladderwort, forget-me-nots, birds-foot-trefoil, campion and endless grasses. I mean, *come on*. Half the time, I had no idea what plants I was looking at, which ones ran through my fingers or brushed my legs, but that didn't matter.

The Devil's-bit scabious is the eminent plant here. In mad, brilliant irreverence, it's also known as 'bobby bright buttons' – a somewhat predictable move from botany (if we just revisit the plant names mentioned above ...). Back in times of scabies and bubonic plague, Devil's-bit was included in the treatment of skin sores – 'scabious' stems from the Latin word *scabere* or 'to scratch'. Legend has it that the Devil bit off the plant's roots in a fit of rage. Anyway, it was sadly not yet in flower when I was there (they usually flower from July). Still, I know it will become a pastel-purple, perfectly spherical jewel. Fitting, of course, for a flower so charming to be wedded to our marsh fritillary, as its larval food plant – the crucial piece in their puzzle. Put Devil's-bit scabious and marsh fritillary together, and they are the power couple of moorland and mire, the equation that solves itself.

Like our marsh fritillary, this plant loves a damp meadow. It quite literally lives for it. It is also partial to a hedgerow, rocky grassland, calcareous grassland, mountain slope and stream bank, affording it a much healthier allotment across the UK. A member of the 'teasel' family, Devil's-bit scabious famously drips in pollen and nectar. Even so, I believe it was its padded leaves and pincushion flower head that drew the marsh fritillary.

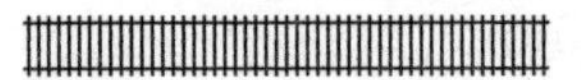

You may recall the joy of the childhood sleepover. The illogical thrill of spending the night on the floor, in a bag, eating different food at weird times and being told off by different parents. Of synching up sleeping bags right up to the top and worming around the room in hysterical shrieks. Well, oddly enough, butterflies do this – in their own way, of course – because a mass web of hundreds of caterpillars jostling for leafy floor space is the route to adulthood for the marsh fritillary. And if you're interested, the Glanville fritillary also develops in this way.

Come July, the first instar hatch. The caterpillars are small and vulnerable, a blur of beige. They remind me of mini versions of retro Smith's salt and vinegar Chipsticks spooned into the protective web of a sleeping bag that glues the leaves of the Devil's-bit scabious together. An afterparty of larvae communing around a leafy table. The bundles of cells I mentioned earlier are fittingly called 'imaginal discs' and are delayed in realising their destiny by being regularly bathed in slurries of a juvenile hormone, which stops the caterpillar from growing up too fast. Think of it as working oppositely to testosterone in a 15-year-old boy suffering from the raging horn. In the teen boy's case, a tide of hormones *promotes* maturation: terminal moods and hair erupting in unfortunate locations in their segue to manhood. Whereas in the butterfly, the hormones racing around the young larva's body serve to *stall* that journey to adulthood – just for a while. Sure, the caterpillar eats, and its internal organs and surrounding muscles grow, but the imaginal discs keep it stunted for a bit. It remains a moving, eating, breathing embryo – pretty much through all six stages. Mildly disturbing, but we'll go with it.

Instar two. The caterpillar matures. A 'moulting' hormone flushes through, triggering the shedding of its skin – yet the juvenile hormone continues to work to keep it a caterpillar. With independence creeping ever closer, instar two is less cul-de-sac sleepover and more 'gap-yah' hostel dorm. The caterpillars have darkened into a tan colour and busily spun a new web around the Devil's-bit scabious, all the while feeding en masse in their foliate canteen. Another dose of the hormonal cocktail darkens the larvae into the third instar, as they busy themselves building an even denser, more concealed web of refuge to help them brave hibernation.

But before they brace for winter, they must moult into the fourth. The timing of this varies from early to late September, depending on the amount of sun and warmth on offer. Looks-wise, our soul-searching traveller has bid *adieu!* to adolescence as an ebony beauty. Jet-black spikes adorn its fourth self to deflect passing birds. The web acts as a kind of hammock. Although less conspicuous now, this silk web-caterpillar combo resembles a black hairball in the grass. Upon closer inspection, however, you'll notice a black, writhing mass of caterpillars absorbing all available heat into late winter. They huddle like students in hats and hoodies at their desks, competing for the last of the hot water and fending off the heating bill.

'It's just *astonishing*', remarked Caroline. 'This tiny mass of two to three hundred caterpillars surviving through winter. While we're all in our heated homes, they've had floods, frosts, snow, gales and all sorts hurled at them.'

When they've made it through the worst of winter, in swings instar number five. Creative differences have broken up the band, and now they forge their separate ways, eager to capitalise on their new solo careers. These caterpillars are still black and still drugged up with hormones, but now they've sexed up a notch and have white polka dots glittering across their bodies. Easy to spot in soggy vegetation, they've left their webs to focus seriously on their new roles as eating machines. Gorging many times their bodyweight in mulched Devil's-bit scabious, it's them against the world.

Caroline loves it. 'You go out on those crisp, sunny spring days, and you find this tiny group of black clustered caterpillars basking in the weak sunshine, and you think – *wow!* There are marsh fritillaries, and they've survived through winter! When you really think about what they've gone through to get to this point …'

April heralds the arrival of the sixth and final instar, a now fully formed, mature, ambitious larva. Ready for some serious adulting. The white speckles are even more sensational. At this stage, we are reminded of the #basic biological urge that drives this unique succession of transition and growth in butterflies and moths. Fundamental, prosaic, innate. After five waves of two hormones at play, juvenile hormone relinquishes its grip on the imaginal discs, instructing them to grow into their specific body parts: the tip of a leg, the corner of a hindwing, the base of an antenna. A final surge of moulting hormone plunges the caterpillar into its white tiger-like chrysalis peppered with knobs and bristles. I'll be honest – if you're looking for magic in this world (aren't we all?) you'll find it in the soupy, embryonic world inside the chrysalis. Our beautiful ebony instar six is mostly recycled into the adult features of the butterfly, tessellating, meshing and conspiring under the canopy of its shell as the imaginal discs start to form a more coherent sentence.

Once the juvenile hormone has had its heyday of suppressing development and is all but disappeared, it's time for metamorphosis. We have now waltzed a full circle and returned to our adult. Evolution's prodigy. Ready to mate, disperse, lay eggs and live a private life. *A Year on the Mire* – the Lepidoptera[3] soap opera on demand. Enchanté.

Right. Let's talk timetables: colour codes, symbolic keys, meaningless subtitles – the lot. As we now know, the life cycle of butterflies and moths is so fastidious and regimented that unless everything happens in the correct order – and in the right place at the right time – it goes arse over tit. If you think you have an impressive five-year plan laid out to perfection, you're about to be trumped. Egg-laying, plants flowering, leaves unfurling, leaves falling … my friend,

[3]The fancy word to describe butterflies and moths (keep a note for your next pub quiz).

nature has had a plan since the dawn of time. The study of all this is called 'phenology', but all you need to worry about is that more and more evidence shows that animals and plants are under pressure to change their 'timetables' in response to climatic stresses. Mismatches in food supply and demand alone would plunge me into a downward spiral of despair. And the thought of the marsh fritillary experiencing such malady leaves a bad taste in my mouth.

According to a few researchers in the Department of Zoology at the University of Oxford, plants and animals are likely to respond to 'climate' in three ways: by shifting their distribution to find 'climate space'; by staying put and adapting 'in situ'; or by adjusting their life cycles. Think of it as a shaky climate causing a jigsaw to misalign. Trying to stick to the plan nowadays is like starting the puzzle with all the middle pieces – a massive effort, and I imagine that species would rather not have to deal with all that admin.

The relationship between butterfly emergence and air temperature is a complicated one, for a good reason. Butterflies reflect the health of a habitat and, with their tight schedules and short generation period, they are sensitive to change – being among the first group of animals to show serious vulnerability to hidden foes like temperature variation. Extensive studies looking at butterfly collections since 1970 at the Natural History Museum found that for every one degree (Celsius) rise in spring temperatures, 92 per cent of the 51 species of butterfly studied responded by emerging one to nine days *earlier*. Case in point, the fine weather of spring 2020 saw the earliest average sightings of butterflies for the past 20 years. Perhaps you noticed?

'There's a common misconception about "this sunny weather" – that rain is for ducks and sunshine is for butterflies,' said Caroline from Butterfly Conservation. 'But we can't afford to think like that any more – it's so much more complex.' True, and when you are fussy enough to rely on only one food source, and the nectar sources aren't there when you dare to arrive early – you're screwed, mate. We also need to talk about a wasp.

Around 80 per cent of butterfly species are prone to attacks by wasps and flies, in a concept known as 'parasitism'. You may be pleased to know that these are not the wasps and flies you are already swatting away. No, these are a totally different genre. Attacks are a gruesome affair and the kind of bedtime story that would deter the average person from a smooth journey into sleep. For a parasitic insect, survival directly depends on the death of its host, following a slick: 'hijack > eat > grow > escape > repeat' programme.

Visits by certain wasps of the genus *Cotesia* are what plague our marsh fritillary, as these wasps have an exclusive appetite for its pretty caterpillars. Roughly three generations of *Cotesia* wasp are produced per generation of marsh fritillary. But try not to see these wasps as a wave of murderers. Instead, think of parasitism as the Neighbourhood Watch of a habitat, structuring ecological communities and policing populations. Obnoxious? Yes. Necessary? Absolutely. In fact, their critical role as 'bio-control agents' in agriculture has become a convenient alibi for their grim behaviour. Helpfully, butterflies reproduce quickly, so any dent in the population from a particularly potent wasp attack can typically be recovered the following year (i.e. in a world where climate change isn't a concern and everything is wonderful).

'Parasitism is *only* a problem if the habitat doesn't have enough of a network to support such complex dynamics – it's a really important part of a robust ecological network.' Rachel Jones, senior ecologist for Butterfly Conservation and PhD researcher at the University of Exeter, confirmed the hunch that we humans tend to fear what we do not understand. We lash out and act up. We make unreasonable assertions. A growing body of (albeit uncertain) evidence suggests that a warmer world may destabilise this crucial interaction between marsh fritillary and *Cotesia* – that the absence of this dyad might unleash a wildfire of our orange marsh fritillary. Not what I was expecting, I'll tell you.

A 2010 study at the University of Oxford found a muddy relationship with weather. Although weather is unlikely to dictate the health of marsh fritillary populations to the extent of, say, habitat quality and connectivity, the authors suggested that sunnier springs could cause the marsh fritillary to emerge first and *outrun* the wasp. Similar work in Cumbria modelled how, without the supervision of the parasitoid, marsh fritillary populations will expand, run out of Devil's-bit scabious, starve, and local extinctions will increase. In short, the loss of this essential population regulator could (in theory) hasten the butterfly towards extinction. Bibbidi-bobbidi-bloody-boo.[4]

We would be unwise to forget that this wasp has persisted for a reason. Like all those in middle management, these wasps are in a constant state of flux, flip-flopping between being followers and leaders as they urge their haphazard habitat towards a state of balance. As you might expect, a large and well-connected habitat will insulate marsh fritillary colonies against extreme attacks from *Cotesia* but, crucially, will still allow the wasps to play the part of the essential bio-control buffer in maintaining sustainable populations. It's that classic predator/prey anecdote where each species is each other's alibi for life. No amount of data can determine what will happen to this unique, co-evolved dynamic in a warmer world. Contemplating what might happen does, however, bring forth a glaringly obvious conclusion: that nature is a lot more bloody complicated than we thought.

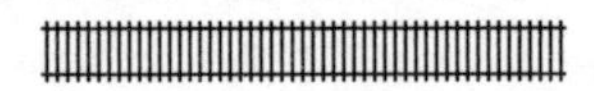

[4]Written in 1948, 'The Magic Song' was first heard in the 1950 film of *Cinderella*.

Back in the heady noughties of 2006, a study by the University of Exeter predicted that the marsh fritillary would become extinct across most of Britain by 2020 if conditions didn't change. As I write, it is 2020, and I can confirm that the species is not yet extinct from most areas – but remains one of the UK's fastest declining species. Yes, 14 years down the line, change has arrived. But it is the selfie, Hinge, Prime delivery, Uber, pandemic kind, you know? We haven't yet decided to sell up and align with the greener way of living. That would surely be far too radical.

Ironically, the marsh fritillary was almost born to die. 'It's a classic metapopulation[5] structure,' Caroline told me. 'This butterfly blinks in and out of extinction over time, even within connected habitat.' Oh dear. If this 'blinking' is as frenzied as that of a certain, rather creepy IT teacher at school, then I'm afraid only God can save the marsh fritillary. Far from anything new, toying with death is part of the narrative tension that writes the marsh fritillary into freedom – conquering new areas and developing sturdier colonies. One year, a colony could collapse when changes in grazing reduce the availability of Devil's-bit scabious. The next, it may be hammered by a demented weather event.

But, as Caroline said, 'In the grand scheme of things, it doesn't matter – because in an ideal world you'll have colonies dotted throughout the landscape and they'll easily recolonise – you will then have a landscape at equilibrium.' Striving for 'equilibrium', of course, being the collective itch across life on Earth. From cell division to circadian rhythms to harmony within the home, we only know we need it after it's gone. 'Boom and bust is *part* of nature,' Caroline stressed, 'but if it happens in a rubbish habitat,

[5] 'Metapopulation' simply means a group of populations of the same species – separated by space. For example, London and its boroughs house a metapopulation of humans (sounds grim, actually).

the butterflies have a slim chance of survival in the long term.' Oscillate in lush lowland grassland – even drier chalk grassland with pleasing foliage corridors – and the boom and bust should be as trivial as an unexpected item in the bagging area: short-term rage, long-term truce.

Please note that a marsh fritillary wants to be *on the move*. Not a ridiculous cross-continent migration like the painted lady,[6] but some localised fidget is to be expected. A 'dispersal' gene is programmed into fritillary DNA – but even a hedge can be a barrier for social butterfly mobility, which is pretty unfair. Smaller, more isolated patches render the innate urge to 'move' redundant – the next nearest site is too far away. With winter flooding increasing and English summer temperatures becoming hot enough to fry an egg on your face, these events are already contributing to low recovery rates and delaying colonisation across the Bodmin populations, even when the habitat is fritillary-friendly.

We can optimise connectivity by creating clear flight paths between a network of good patches. 'You could have chalk grasslands and farmland in between,' admitted Rachel, 'but that farmland between habitat patches could be *softened* with margins, buffer strips and nectar sites – it's not as hostile as a huge field of barley or wheat.' Trouble is, our landscapes are changing at a pace with which nature struggles to keep up. Our hunger for development is not giving nature enough time to respond. Not ideal when we have created a world that craves instant response. We're not giving this butterfly long enough to reply.

Even in places (like Bodmin) where the habitat has altered for the better, an undercurrent of angst means that

[6]An annual summer immigrant to the UK, these butterflies undertake one of the most astonishing migrations. Cruising at speeds of 30mph, painted ladies clock a 9,000-mile round trip from tropical Africa to the Arctic Circle. It is thought that migratory butterflies orientate their flight using the sun as a compass.

fragile species like the marsh fritillary remain in a state of 'extinction debt' – where they are, quite literally, indebted to extinction and owe it their life. This term soberly refers to a 'delayed' extinction event, where a series of dire circumstances *should* have led to species disappearing – but haven't. Thus, the species (i.e. marsh fritillary) exists in a sorry state of limbo, waiting for the inevitable. 'There is a balance between a watchful eye and letting habitat become totally reclaimed.' said Caroline. 'It's a classic "Goldilocks" scenario of habitat suitability.' Makes sense. Such logic is applied elsewhere, from economics to the habitable zone around a star. Balance *must* exist somewhere between the sweaty human grip of nature's scruff and total abandon.

A good example lies in the choices we make for the land. Devil's-bit scabious doesn't do itself any favours by being a tasty plant for sheep (the most aggressive grazers). Historically, large areas of sheep grazing in the South West (and Scandinavia, for that matter) have driven a gross loss of plant structure – and food sources for the marsh fritillary. Not only that, but overgrazing can expose a more significant issue. 'It can make a habitat prone to drought,' mused Rachel. 'We don't have enough data to know whether the Devil's-bit scabious will be able to withstand it, as populations can tolerate occasional drought. But it's the frequency and intensity of recent droughts that could be the killer.'

But, weirdly, Caroline explained how a total abandonment of grazing could be just as detrimental. 'We need grazing to maintain the mix of short to long sward that Devil's-bit scabious and the butterfly need – but when it's too heavy, that variation is lost.' Gotcha, Goldilocks. Populations of marsh fritillaries in areas of Dorset were up fivefold where grazing was 'recently abandoned', i.e. not mown to the ground, but also not wild and disorderly. Much like a decent haircut, it is the *height* of grassland that is key. When you're a butterfly with a diameter that spans the average length of the human little finger – and you're striving for a mate and a place to lay your eggs – an ungrazed swathe of grassland is a demoralising fortress. Equally, however, you don't want to have to fly long distances over barren lawns bearing your heavy, eggy burden. So, Caroline's team found that a 'lightly grazed' area yielded a vegetation height of 5–20cm, which seemed to produce lovely, connected plots of Devil's-bit scabious. Scales a-balanced.

For us, 'connectivity' simply requires a virtual network bot to guide you, a sweating you to plug some wires into a router. But for the marsh fritillary? Simple recruitment of gentle grazers and a healthy willingness among humans to collaborate with each other (and with ecology, obvs) will suffice. 'The critical need for survival is good quality habitat, followed by large patches of good quality habitat, followed by well-connected land,' Caroline explained, stressing how important the latter element is for enabling colonisation in the event of local extinctions. 'Populations can persist without mixing, but if local extinction occurs, re-colonisation is impossible.'

When I saw them, I was mentally composing a disgruntled email to lovely Cornwall Butterfly Conservation's Jo, huffing at the so-called 'guaranteed hotspot'. Then, not one, but two flickers of orange confetti scattered in the wind against a blackened Bodmin sky. A fragile carousel, dancing with a higher purpose. I was lost to the moment. A glimpse so fleeting, I wondered whether it really happened. Still do. Ask any butterfly fan, though, and the consensus is that *you just know* when you've seen a marsh fritillary. Modern society has primed us to view every novel encounter with nature as something to feel deeply moved by, as though we have uncovered some fundamental truth that has evaded seasoned naturalists for centuries. When in fact, it's merely a chance (fine, a slightly engineered) rendezvous with something that is not another person.

The pair I saw disappeared as fleetingly as they had arrived, behind the granite fallout from the tor – seeking asylum in a patch of grass that looked as though someone had scribbled it with a biro. Perhaps they had gone to mate, lay eggs, conclude their chapter? Such is their commitment to the cause. Or maybe this *pas de deux* was their finale?

The laden calm before the weather shifted blanketed Breney Common. Anticipation was replaced with a kind of tipsy placidity. Finding a slab of rock on which to sit and eat my apple, I enjoyed that rare pleasure of a total absence of thought.

There's a nostalgia that comes with pursuing a butterfly. The simplicity of a day out with a packed lunch and a wholesome can-do attitude revives a time of ease and contentment when decisions and *things* were less of a labyrinth and more of a leafy avenue with clear views into the distance. I floated this idea past Rachel and was comforted with the notion that I wasn't being a total salad. She told me the marsh fritillary is just associated with a lovely landscape. That it was the first butterfly she had worked on. Recalling the childlike joy she felt upon seeing it for the first time, 'It just symbolises a landscape that *was* – and I don't think people realise that.'

A day like the one I spent on Helman Tor is childhood revisited. Sweet as summer rain. Fabricating memories, maybe, but these are surely the ones to bookmark? Gorse brushed against my backpack, sounding like hair might if it swept across canvas. I wondered – if an artist had painted this entire scene, where would they place me in the landscape? Would I belong in it at all? I rolled a spiral of fern into my fingers and ground it into some kind of pesto, closed my eyes and smelled it.

In more ways than one, Bodmin is the overlooked success story. Neighbouring Dartmoor has a similar resumé and has provided valuable lessons that inspired later work on Bodmin. A long-term lack of grazing led to a general vibe of disarray, overpowering Devil's-bit scabious and leaving Dartmoor's population of marsh fritillaries isolated and pissed off. In the spirit of not-letting-it-all-go-to-shit, Butterfly Conservation launched the Two Moors Threatened Butterfly Project, placing actual humans (rogue!) on the ground to increase connectivity, add grazers, train staff and conduct guided walks. In a wild, brilliant example of archaic abandon, landowners and farmers were supported and educated by agri-environment schemes on how to cultivate *in favour* of the butterfly, at no cost to their productivity. Unsurprisingly, connectivity quickly increased. Flight paths of the butterfly nearly halved to 260 metres. Larval food webs rose by 1,080 per cent, and 71 per cent of sites across Dartmoor are now actively managed for the marsh fritillary, earning it a well-deserved position as a national stronghold for our little *aurinia*.

Back to Bodmin, and between 2017–2020, Butterfly Conservation launched All the Moor Butterflies. Twenty-nine new marsh fritillary sites were restored thanks to 165 hectares of farm-specific habitat improvement – a roaring success. Traditional practices of 'swaling' (controlled burning[7]) and rotational cattle grazing are optimal for maintaining the numbers here in Cornwall – repaying the

[7] Not to be confused with other South-Westerly activities such as 'welly wanging' or 'wassailing'.

extinction debt accumulated across these Cornish moors. In short – conservation *can* work. And we should stop being so surprised by this.

In an age where *Build, Build, Build!* and *Shop for Britain!* seem the societal hymns, small and connected patches jammed with butterfly food plants is all we ask for. Keeping ironies intact, Britain's future-proofing strategy has so far consisted of: debate > million-pound flood-defences > commission Progress Report #5,000 > debate > design overpriced electric transport > convene Nothing is Getting Done Committee > fashion 'net-zero' rhetoric > debate > plant some trees. When all we need are stepping stones of quality habitat to lead *any* fragile species away from a battering – us included. We seem to forget that it is in *our* best interest to preserve the habitat of the marsh fritillary. Sure, we would sorely miss it as an upland jewel, but its absence would be symptomatic of a landscape in failing health. A vaccination for that disease will never exist. Sorry.

'*Everything* is interlinked,' said Rachel, '*that's* what we keep forgetting.' This butterfly is teaching us a lesson. Quite right. Coaxing populations elsewhere across the country has had mixed results, but the beauty is we already have pockets of colonies sitting on habitats that we would be wise to preserve. It's a question of capitalising on existing populations, making the habitat as rich a tapestry as possible, so that if climate change does give us an almighty kick up the arse, the resilience is there to fend off the challenge and keep these butterflies in play. As Rachel pointed out, 'creating strong networks of habitat will bolster those existing populations. Sustained through the legacies of projects and management, the butterflies will be able to track naturally.' That sounds good.

There is a chance that the marsh fritillary's curious ability to occupy different niches around the world might stand it in good stead as our climate becomes more Mediterranean.

In southern Spain, it swaps moist tussocks for woodland and Devil's-bit scabious for honeysuckle. Caroline reflected, 'In 50 years, the marsh fritillary in England might be in the woods eating honeysuckle, but we must maintain populations in the present, so future generations have the chance to adapt. Who knows … but it's a thought.'

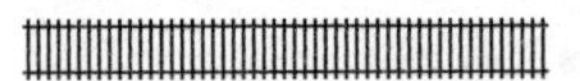

Rain finally announced my cue to leave. The kind of rain that is exclusive to a British moorland. Despite the morbid reality of being drenched, cold and hungry, the freedom of being outdoors and powering my way home felt as physically freeing as taking off a bra at the end of a long day. My hands were frozen and weirdly coloured, barely able to shift gears, let alone use the brakes. I had never been wetter. The moment had all the components of a scenario that should sap the soul – but, weirdly, I felt grand. Having gone far too well anyway, it made sense that I had a full-bodied experience right to the day's end.

I reckon the marsh fritillary is the perfect metaphor for change. It lives only as a result of extraordinary transfiguration. But I also reckon we are very quickly writing it out of its own story. Like a teacher aggressively circling 'marsh fritillary' with a red pen, commenting 'does this add anything???' – before we've taken the time to consider whether *we* could be the ones who need the edit.

I made the most of the train journey back to Exeter, stripping off clothes for the second time that day and transforming the empty carriage into a launderette. I felt shaky. Hating myself for what swiftly emerged as appalling catering planning when all available food opportunities while travelling were on pause. I was branded by the day; chain grease tracked up my calf like some heavy-metal tattoo. Changing trains at Plymouth, I latched on to the only available vending machine like a moth to a lamp, buying two Twirls and some yoghurt raisins, and waited on the platform, damp and shivering in a sugar-trance.

Naturally, not a drop of rain had fallen beyond Bodmin. I arrived home as moist as Breney Common itself. *Eurgh. Good practice, though*, I thought, watching tiny bubbles burst around the seams of my sodden trainers. My next trip promised even more water. To Wales.

CHAPTER TWO

Harbour Porpoise

I've been in an emotional relationship with Wales for about 15 years now. Moving from the US state of Georgia – I was three and my brother was six – our former-naval-officer parents were keen to expand our world. Looking back, I love them for it. We are the family that takes the stairs. Much to the alarm of some peers, our summer holidays featured tents, zips, walking boots and misty summits. Also, soup. Key highlights of our Welsh camping series include repairing a broken tent pole's elastic with strands of my hair, and employing all pots and pans to save a leaking tent during a particularly miserable attack of sinusitis. Despite dearest Mum vowing never to camp again, they really were good times.

It was a Tuesday. It felt good to have a train to catch on a Tuesday and for that train to be heading (eventually) to Pembrokeshire – the jade bracelet of Wales I had heard so much about but never visited.

A far cry from squatting in the tussocks of Breney Common, my route headed north before bending west – the plan being to spend the rest of the week feeling up the Welsh coastline and (hopefully) finishing somewhere near Snowdonia. I was running unfashionably late and caught my bike in the automatic doors separating the train carriages, an oversight caused by missing the designated bike carriage. After failed attempts at elbowing buttons and getting very cross – the doors suddenly released. My handlebars flipped the front wheel around, smacking into a passenger's leg. Apologising profusely for failing to have the situation in hand, I was met with a glare from said passenger that could have shattered time itself. A confident start.

I had fallen into a simple routine. Board train, wrestle with bike, savour the journey and hope specific animals may dare to show up. My bike crashed loudly against the racks at every jostle, adding a metallic edge to a green and pleasant ride as we raced through the Exe Valley into Somerset. Dairy cows drifted between fields, and I thought of beavers at every fallen branch or cluster of sticks as we followed the river. I found myself slumped diagonally across a window seat with my legs wide apart like some guy at a bar, drinking in the blur outside. I was very content. This train was the line home during my university years in Bristol. During those brilliant, bizarre years, I would experience a genuine buoyancy knowing that the sea, the moors, my own bed (also laundry services, proper meals) and the absence of an alcoholic agenda were waiting just over an hour away. This train line has accommodated most if not all versions of Sophie over the past five years.

I was like many children – horses, dolphins and the sea ruled my world. During the early years of primary school, our teachers asked us what we wanted to be when we grew up. We are so much more at ease with ambition at that age. I wrote 'underwater photographer' and, in later years, 'pilot', before climaxing during an especially awkward phase of gelling my hair and being desperate to legally change my name to Janet (genuinely) when I put 'guinea pig breeder'. My bedroom rocked a cetacean vibe right into my teens. I branded every available surface with fin and fluke. Arm, napkin, exercise book, toilet roll – all would be adorned with the same underwater scene: novelty wave at the top, seaweed and rocks at the bottom.

My ocean was busy with life at all levels: dolphins, whales, schools of fish, an octopus, starfish, sceptical rays; all surrounded by tiny circles, which I think were meant to be bubbles (from some unchartered hydrothermal vent, perhaps?) A scuba diver surveyed the scene, always from the right-hand side. She was never that much smaller than the giant whale coming up for air next to her because, in this world, we exist on a level.

But it never occurred to me to include a harbour porpoise in this medley. It just wasn't on my mind in the same way that dolphins and other whales were. We didn't learn about porpoises in school. They did not feature in any bedtime stories. A real shame, as they're one of the ocean's smallest mammals and Britain's smallest cetacean (the group of aquatic mammals that includes whales and dolphins).

No more than 2 metres long, a harbour porpoise weighs about the same as a human teenager. It's a similar blue-grey colour to the bottlenose dolphins we know, but with a more rounded, blunt head. Like me during the Janet/guinea pig breeder phase, the harbour porpoise is famously shy – most often seen travelling in groups of two to three around coastal habitats. Its apparent omission from pop culture and listing as a highly protected 'priority species' for UK waters is what drew me to its story, along with the fact its Latin name *Phocoena phocoena* means 'pig fish'. Its characteristic 'chuff' as it surfaces for air inspired its fond nickname of the 'puffing pig'.

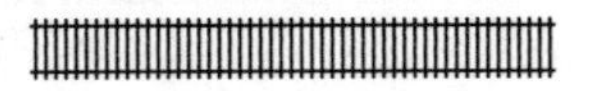

Our grasp of 'the ocean' has developed at a similar pace to our understanding of the human brain. Both remain devoid of, yet are the compass to, a richer understanding of the planet and our place in it. Understanding the ocean is understanding ourselves and our survival. With its immense peaks and troughs, the deep sea could be an analogue to the abyss of the human subconscious – the unmapped mountain ranges of our mind. We have explored less than 5 per cent of the ocean and charted only 80 out of a possible 180 areas of the human cerebral cortex. But we know we are smart. We've innovated, flexed our ego, upgraded – but to what end?

Cast your mind back to 180,000 years ago. You are in a dim, cool cave off South Africa's Pinnacle Point coast, and it's dinner time. Last night it was *Hypoxis*, yellow stargrass roasted in ashes – again. Wooden sticks litter the firepit, your body aches with the labour of uprooting these fibrous rhizomes from the hillside. A bowlful of its white flesh offers about the same energy as a very large banana, tasting sweet if a little nutty. But hey, tonight is different. For tonight colludes an affair with barnacle, giant periwinkle and whelk – an amuse-bouche of crustacean and gastropod. At last, the language of food adopts a new meaning. The prospect of this new, fishy nutrition promises to make you nimble and strong. Tonight weds man and sea.

This version of events recounts the supposed dawn of the human obsession with fish. It was 2007 when a group of paleoanthropologists from Arizona State University led an excavation of a high, dry cave at Pinnacle Point. A decent portion of shellfish was uncovered, dating back to nearly 180,000 years ago. Not quite fish and chips, but the discovery ebbed the earliest known human seafood meal by 40,000 years. The researchers identified some barnacles as being harvested from whales. Now I can't imagine that empire building and trade routes were on the average hominid mind back then, but somehow it all started to happen soon after. Curiosity killed the ocean. Jonah, Medusa, sirens, superstition, merpeople, Davy Jones, the omen of a cormorant – even I still mumble, 'Red sky at night, sailor's delight' (cheers, Pops!) – the human romance with the waves is deeply woven into the tapestry of civilisation. And yet we've used and abused. Drilled, overfished, polluted. Our current relationship status: as complicated as a sushi conveyor belt.

Highways of buddleia[8] grinned purple for miles along the edge of the track, assuming identity in abandon. Swansea Bay opened like a glossy magazine to my left. For more than 60 years, Port Talbot has entwined with the steel industry, employing up to 20,000 workers in the 1960s. Yet falling demand and brutal Chinese competition have rendered this historic site nearly obsolete. Tiny terraced houses remained braced against the rest of Swansea Bay. Industrial, ebony shafts rose above it all, like giant cigarettes.

Swansea station was sunny and predictable. A high-rise block of flats was mid-facelift, masked in a lilac construction plastic with two enormous cranes sticking up like antennae. Five herring gulls circled noisily above. This did not feel like Wales. When I'm in Wales, I exist (on the whole) in perpetual damp. I am undoubtedly also experiencing wonderful, carefree times, of course, truly wild – some of the best. But inevitably damp and under canvas. The sun gained confidence, and I thought of the sea. I could almost smell it.

Through the train window, the open sea collided with the marshes and massive estuary of the River Loughor, blending with the 450-acre (182-hectare) birder wonderland

[8] Sometimes called Britain's 'national flower', this invasive plant was introduced in the late nineteenth century from China and Japan and has a knack for growing in the most awkward places, spreading rife across railways and buildings.

of Llanelli Wetland Centre. Beaches tinged gold; headlands flushed green, the Gower Peninsula unfurled like a new leaf as we raced west. On a map, the outline of the peninsula assumes a reasonable bellend, the very tip beyond the headland wonderfully named 'Worm's Head'. But (phallus aside), here lies a village of world-class beaches, saltwater marshes, heathland, limestone grassland, ancient woodland, sand dunes, rocky shore, Iron Age forts, medieval castles, above-average surf and well-above-average coastal path. This 19-mile stretch of Arcadia was Britain's first Area of Outstanding Natural Beauty (AONB) and rightfully so.

Carmarthen Bay (Bae Caerfyrddin) reminded me of Cornwall's Camel Estuary near Padstow, both a similar weight of golds and teals, while waders picked their way along the shore like fossil hunters. Grand white houses and cosy towns. A golf course had clumps of yellow ragwort dotting the fringes of polished lawns. Beaming at their permission to stay, I was just thinking how, if viewed from above, the ragwort would look like giant dandelions when a 22-spot ladybird climbed onto the table and mapped the length of my index finger. One of three types of yellow ladybird in the UK, it marched along my finger eagerly like a tiny cadet, surveying this fleshy mass for food, shelter or mating opportunities, legs and antennae reading the scene like morse code. Upon realising the futility of this operation, it moved on elsewhere, and I lost sight of it.

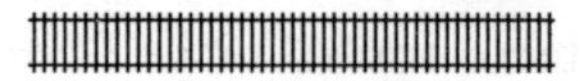

I arrived at Fishguard & Goodwick station on a day that forecast serious warmth. Desperate to stretch my legs, I mounted my bike and freewheeled towards the sea, arriving almost immediately at my meeting place. The coast path above the beach stitched the headland like a suture. Oystercatchers triumphed over traffic. I had a couple of hours to kill before I was due to meet with Holly, a marine biologist and project officer for conservation charity Sea Trust Wales. I bought a sandwich and settled on a rock overlooking Goodwick Sands beach. Sitting high enough to swing my legs from the wall, I watched a young boy. It took me a moment to realise that he was singing to himself as he turned over rocks, prodding the mud with his fingers, bringing them up to his nose for a good sniff. He busied himself assembling his castle of riches on the edge of the tide, with the care of someone beyond his years.

I had told Holly to look out for an unkempt trio of girl, bike and backpack in the car park. In a red car, she pulled up beside the Ocean Lab. Holly is the kind of person you realise you like very much, even before she says hello. Bare-faced, bright and with a soft Surrey accent, there was a certainty about her that soothed me. Being just a week into the easing of national lockdown restrictions, Wales remained hushed, and the Ocean Lab and its public aquarium were closed. Dim lighting filled empty tanks – the contrast quite gloomy. 'We had quite the task of gathering all the species from tanks and releasing them back into the sea when lockdown started!' laughed Holly. 'Why should they stay indoors just because we had to?' We like Holly.

Leading up to that day, Holly had teased that on a 'good day' with the right conditions, harbour porpoises could be seen from the coast path at Strumble Head. She had invited me to join her on one of her daily porpoise surveys, and I played excitable intern as she handed me a pair of large binoculars and led the way. In pre-pandemic times, car-loads of Sea Trust Wales staff and volunteers would set up stations at four different local sites each day – Strumble Head, Fishguard Harbour, Ramsay Sound and Pen Anglas – to collect vital photographic data as part of the Porpoise Photo-ID Project (try saying that after a few). But, as society stopped, so did the surveys. 'It was really painful,'

Holly admitted. 'We've surveyed consistently for the past three years, and we *need* this data. So, as the restrictions have lifted, we locals have resumed our surveys – thank God.'

The lanes leading up to the headland had echoes of Dartmoor. Tiny stone holiday cottages pocked the hills. It was as though the North Devon and Cornish coastlines had conspired (with the blessing of Bodmin Moor) to have a secret child. A ferry chugged away to Ireland. An ellipsis of gannets headed home to Grassholm Island. There was so much sky I felt like I was walking into a fish-eye lens – the further my entry, the wider it became. We settled on a spongy area of grass on a flat ledge away from the coast path. Mustard-coloured lichen whispered across the rocks. Pollinators bustled about their coastal abode. The air was fresh, seasoned like salted butter – and I wanted to spread it all on a bagel and eat it. 'Special place, isn't it,' smiled Holly. 'There's a peregrine nesting in the rocks over there. We even get adders – and we're just below the car park! Every time I come here, there's something different.'

A man with an enormous camera lens beamed at us as we arrived at our survey spot. 'Alright, Rob?' Holly greeted him like she was pulling up a barstool – I clocked 'the nod' of fellow nature lovers. 'Any luck?' she asked.

'Gah!' Rob cleared his throat behind an equally enormous beard, 'Just the usual I'd say!' in a thick Welsh lilt. 'Won't see any porpoises up here 'til high tide now, but she's a beauty, mind!' He grinned at the sky, his beard wobbling with the wind. Hoisting his tripod on his shoulder, off he went up the hill.

I was on the cusp of quizzing Holly about porpoises and such when our eardrums were scoured like a pan. Had this happened the year before, this sound could have passed off as a crow with severe laryngitis. However, having since been blessed with the friendship and know-how of some talented bird-y people, before I knew what was happening, I shouted

'*CHOOUUUGHS!*' (pronounced 'chuffs'). Two large, ebony-black birds rolled in front of us, jumping a jive on the rocks. Pepper-red legs and a long, curved bill to match. These birds have mastered flight and survival beyond all the corvid[9] cousins. It was between 1860 and 1900 when people suddenly realised that they'd very much like to collect the increasingly rare chough's eggs and join the 'collector' hype of the Victorian era, please. Although the study of egg collections now enables extensive understanding of the early naturalist expeditions, bird evolution and human culture (the Natural History Museum has archived more than a million individual eggs), this along with the trapping and shooting of choughs plunged this jolly bird into extinction over much of England during the 1900s.

It's been a turbulent ride since, thanks to the intensification of farming and abandonment of grazing, dung and insect food on the coastal slopes it loves. But, thanks to an army of good people, it is now afforded the highest degree of legal protection. Numbers are recovering in most coastal regions of the UK – including here on Strumble Head, where there are two breeding pairs. Watching them flash their wings and land with a bounce, feathered fingers outstretched, these choughs lacked the sinister witch doctor vibes of the jackdaw and assumed a more comedic role. Their unmistakable '*chee-oowwww*' call caught on the breeze as if amused at their recent U-turn of fortune.

I looked across the sea, which yawned ahead of us. Water upwelled into soft little circles called 'boils', like a reluctant jacuzzi. Generally speaking, the more boils, the lower the tide. I lost count of the boils. The tide was very low. Harbour porpoises rarely feed at low tide. More likely, the ones here feed at mid-tide. Holly had no problem reminding me that

[9] Members of the crow family, including crows, jackdaws, rooks, ravens, jays, magpies, Asian treepies, choughs and nutcrackers.

I had picked *the worst* possible time to see harbour porpoises. We laughed for the sake of it – the sort of 'funny ha-ha' type, devoid of all hilarity. I remained optimistic as ever but typically have more luck with pretty much anything else than I do with wildlife performing on-demand.

I had seen harbour porpoises before, though. A few years ago, alone in Falmouth with time to spare, I treated myself to a boat trip. It was one of those days where the Cornish sea and sky adopted the same shade of slate, and the lack of sun flare invited perfect cetacean spotting. A few cormorants later, and the faintest glimmer of a mother and calf wrinkled the surface. As shy as the November sun, their dorsal fins smooth and blunt, their bodies bashful but sturdy.

Often mistaken for each other, the porpoise and the dolphin share many qualities – but bravado is not one. Your average dolphin is like *that guy* in the office. Tailored, clean-shaven, a healthy 6 feet 4 inches. A 10-metre wake of Paco Rabanne. *Loves* a good laugh, babe. Can be an occasional arse. And definitely wouldn't call you back. Whereas our harbour porpoise is the cute guy in the corner who enjoys knitwear and helps clean up after the Christmas party. And only *after* the third pub quiz do you find out that he studied at Harvard, has three PhDs and would most likely make you a mixtape.

Such social reluctance from the harbour porpoise has made them notoriously difficult to study. Despite it being one of the most common and savvy mammals in global seas, we regrettably know more about the anatomy of a Kardashian than we do the nuances that chaperone our porpoises through life. It's nobody's fault – it's just one of those things. And, ironies wonderfully intact, it's this enigma that spurred the idea of a camera becoming one of the porpoise's greatest modern allies.

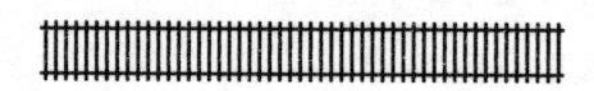

Sea Trust Wales' focus on the harbour porpoise began on a dark February evening in 1996. The huge Liberian oil tanker, the *Sea Empress,* ground to a halt on St Ann's Head rocks in nearby Milford Haven – one of the UK's most frantic and largest energy ports. A gross assault of 72,000 tonnes of crude oil and 480 tonnes of fuel poured into these waters. The 5-mile oil slick polluted 200 kilometres of Pembrokeshire National Park coastline; it remains one of Britain's most serious oil spills. Amid the anguish of rescuing more than 7,000 oil-bound birds, it became apparent that despite large marine mammals remaining relatively untouched by the disaster, there wasn't a vested local authority to respond solely to the conservation of marine wildlife in general. After being born in the spare room of Holly's boss in 2003, 17 years, an Ocean Lab, aquarium and 40 volunteers later, Sea Trust Wales' Porpoise Photo-ID Project remains one of just a handful of organisations around the world collecting data on porpoises in this way.

While recording dangerous and thrilling dives, legendary French explorer Jacques Cousteau revealed the life within the sea to millions from the Second World War onwards. His 1953 book *Silent World* cements his pioneering legacy of underwater photography and adventure. For Cousteau, photography was vital to his reconnaissance. Holly can relate: 'People used to tell me I couldn't photo-ID porpoises – that it was impossible,' she chuckled. 'That's how this project started because volunteers were up here getting photos and sending in their sightings. Like a catalogue! And we've now got 138 porpoises recorded – and counting,' she grinned.

The notion of a 'porpoise catalogue' is one to which I wish to subscribe. Just imagine, every month, this popping through the post box? Perhaps you ferry it back to a corner, don some woollens and hunker down with a tea – ready to scoop the secret lives of the Pembrokeshire porpoises.

This catalogue is the result of one of the first studies of its kind in the world. Arguably the best, it's changing the way that porpoises are studied. Photos of random porpoise heads and fins reveal stories of their families and social structures that were previously a mystery. The 'notches' in dorsal fins are as distinct as a fingerprint – and a photo of this is enough for science.

According to the thousands of photos that Holly and the team have gathered so far, we've got around six key characters at Strumble Head: one incredibly diligent female has had four calves in four years. Same father? Doubt it. Another is a desperate housewife craving empty-nest syndrome, after both calves born between 2017–18 are *still* at home, in a curious demonstration of the 'boomerang generation' (*cough*, cannot relate). Exotic lone travellers occasionally arrive on the scene. Perhaps on a journey to find themselves? Or on a romantic mission to normalise toxic masculinity (male harbour porpoises have been known to emit aggressive click patterns on occasion). Please. It's aching to be adapted into an ITV reality binge.

'People just don't tend to study harbour porpoises in this way. They're small, shy, they don't come out of the water and perform like dolphins, but *here* …' Holly greeted the scene like a master of ceremony, her right arm sweeping the horizon, '… here you can see the porpoises coming in, you can *see* their trajectory and direction of travel, how fast they're going and then you can take photos and see who it is.'

Wonderfully, through Sea Trust Wales, you can now 'Adopt a Porpoise' and be updated with news of calves and the latest gossip. Holly finds that people who approach her are more than happy to be corrected and learn. People forget. And that's OK. Porpoises just aren't *that guy*.

Near where Holly and I stood lies a body of water known as 'The Bitches'. It describes a frenzied composition of immense tidal races, rocks and reefs, creating tides that muscle through narrow channels into Ramsay Sound at speeds of up to 18 knots – around 20 miles per hour. But here's the thing. Whenever I am in a car, travelling at 20 miles per hour, I feel *certain* that my walking pace is faster. It just feels unbelievably slow. Place yourself in a body of water moving at such speed, however, and it is an entirely different story. Caught up in it at the wrong time, with the wrong equipment, and you might find yourself shouting a vocabulary far more colourful than 'bitches'. It is something of a white-water rodeo show and thus is home to the international kayaking championships and many a shipwreck beneath the whirlpools. It's these tidal races that are an essential agent in the harbour porpoises' assignment. The parent current rises up from the south of the Irish Sea, with an offshoot hitting the Llŷn Peninsula off the west of Snowdonia, before ringing back south around Cardigan Bay. It's got plenty of time to gather speed and power to govern these coasts.

Holly traced a line across the sea in front of us, running diagonally like a trouser seam. I squinted my eyes. 'This bit collides with the eddy circling back from Cardigan Bay and acts as a trap for the fish – which is why so many animals come here to feed.' Seals, sunfish; common, Risso's and bottlenose dolphins; as well as minke whales, gannets, guillemots, razorbills, Manx shearwaters, even a novelty killer whale on occasion – the entire aquatic A-list has a role in this exceptional matinee.

Looking below, I willed the swell to gather and sweep a dozen porpoises in its arms. At high tide, the porpoises often leap out of the waves while feeding, and it's possible to get a photo of their entire body. In winter, they surf the waves. Holly was scrolling through photos of such things on her phone. On flat, calm days like this, a glimmer of a fin breaking the surface is all that you are likely to see. Some more boils burped on the surface. The scene was every shade of blue. Any other day, any other endeavour, and it would have been meditative. But the itch of impatience was growing on me like a monster spot. I tried to impress Holly by confidently and – incorrectly – identifying a sunfish below. 'Oh, yeah – we've had loads of compass jellyfish here this year,' she said casually – course you have. I wanted to dissolve and return to the world as a woodlouse. (To be fair, it was pretty far away.)

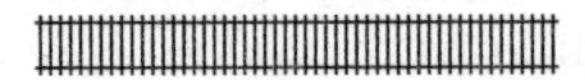

It's not just marine life that has monopolised this unique confusion of tides. The human quest for alternative energy sources has, in recent years, adopted a more fervent approach. In the UK, this is particularly urgent, given that we have legally bound ourselves to become a carbon-neutral economy by 2050. An admirable, if aspirational, idea given that our growth in low-carbon electricity production slowed sharply in 2019. Wind, wave and solar power being the key trio in our royal flush, in what sometimes feels like an infinite global poker game as we race to accomplish carbon neutrality.

Several elements do work in our favour here. More than 6,000 islands form the British Isles. The length of the collective coastline is nearly 20,000 miles (just shy of the circumference of Earth). Our northerly latitude plops us amid a major convergence of air masses – ensuring our weather is good but occasionally poor. Admiral Robert Fitzroy began issuing maritime storm warnings in the late 1860s, founding what would later become the UK Met Office. Rob must have needed another challenge after captaining HMS *Beagle* on Charles Darwin's celebrated circumnavigation in the 1830s. Fitzroy's storm warnings sparked the genesis of today's weather forecasting and, since 1925, the iconic poetry of BBC Radio 4's *Shipping Forecast* remains the longest-running continuous forecast in the world – covering 31 sea areas four times a day. Fishing rigs, cargo ships – vessels of every shape and size – someone, somewhere, relies on this airwave medley to navigate the sandbanks of Viking and Forties, the islands of Utsire and yes, even Fitzroy.

We are a wet, windy, gloriously maritime nation. Renewable energy is by no means the solution to climate change – but it sure works in our favour. And we are good at it. There's just that small problem of *space*. The UK Offshore Wind Sector Deal aims to upgrade to a 'green economy' and power every home in the UK via offshore wind farms by 2030 – four times what it was in 2020. I applaud the ambition. But to put it mildly, the logistics surrounding the construction of renewable energy devices is a pain in the national arse. Let alone the palaver that comes with updating tired old onshore regulations to accommodate these green upgrades. The point is that the area around Strumble Head where I stood with Holly is under *constant assessment* for tidal and wind energy. A box on a spreadsheet somewhere is aching to be ticked.

For Holly, it's only a matter of time before this horizon includes a series of 120-metre high turbines wielding blades of up to 80 metres long. A slight detour might be required for our gannets returning to Grassholm. 'It's the construction of the turbines that cause the most damage,' Holly told me, 'it just displaces the porpoises – shoves them aside for a bit.' But it's the age-old question of short-term impact versus long-term gain. Sacrifice a few species for the good of the many? The irony of a greener solution harming nature cannot be ignored.

This dilemma is the focus of leading research from Swansea University. Dr Hanna Nuuttila is a post-doctorate research fellow at the university. She's one of these amazing women who seem to be jumping all of life's hurdles while studying open ocean ecology, marine mammals and marine renewable energy. Famously, she was also arrested in 2019 for using her body as a blockade on Oxford Circus as part of an Extinction Rebellion protest. She was seven months pregnant at the time. 'I did it for my children and their future,' she told me.

Hanna is part of the team with which Holly and Sea Trust Wales liaise. Using acoustic monitoring devices, she led the first study that recorded porpoises and dolphins throughout the year in Wales – advocating for their protection. I can't remember the last time I felt so impacted by someone's passion for the ocean. She radiated love for her work and energised me about a subject that has a high chance of being boring.

'Marine renewables are a tricky one because a lot of people are very against them,' Hanna explained. 'But I look at the bigger picture and think, "Well, what are the other options for where our energy comes from?"'

Wind energy is costly to harvest and unpredictable in supply. Ten years ago, offshore wind farms were 90 per cent more expensive than fossil fuel generators and 50 per cent more expensive than nuclear. Despite bigger, better turbines reducing the cost of wind power, a significant wad of cash is still needed to meet the government's net-zero carbon target by 2050. Something in the region of a £50 billion investment and one turbine installed every weekday for the next 10 years are the current thoughts so far.

Hanna is one of several researchers maturing our understanding of how our adaptation to climate change affects the ability of the marine world to do the same. In high tidal zones like Strumble Head, there was little to no relevant research on how renewable devices would impact marine mammals. Species that, if threatened, have the power to initiate a cascade of trouble for the rest of the food chain. Hanna seeks to remedy this. In 2017 she wrote an academic paper entitled 'Don't forget the porpoise'. I knew I had come to the right woman.

Harbour porpoises are highly mobile. How far they can travel in a day is still barely known, but judging by their presence across most waters of the northern hemisphere, it seems likely they are very well travelled. Hanna explained that the trouble with animals like this is we may never be able to understand the effects of artificial structures on their lifestyle, 'because they can just choose to bugger off!' Quite right. The proposed turbines and development around nearby areas of the Welsh coast like Ramsey Sound suddenly make Holly's surveys seem like the most important job in the world. Hanna led a report in 2015 that appealed the need for these 'fixed vantage-point surveys' to understand better what rows of steel giants will mean for this part of the British Isles. Taking photos of porpoises and engaging the public with their story has never been more crucial.

In terms of an immediate impact to the porpoises and other wildlife around Strumble Head, it's thought to be similar to those that arise from the exploration for – and extraction of – commercial oil and gas since the 1850s: noise pollution, contamination, changes to water flow and turbidity, changes in food availability and a very real risk of cetaceans colliding with the turbines or increased boat traffic during the building phase, which can last anything between five and seven years.

A German study in 2008 observed the impacts of an offshore wind farm on harbour porpoise behaviour, right in the infant stages of the farm's construction. Horns Rev 2 was the first offshore wind farm in the North Sea, with 91 turbines standing off the coast of Denmark. It honed the technique of 'pile driving' (ouch ...), where enormous devices are plunged into the seabed to lay the foundations for the turbines. Pile driving has been a thing since Roman times and has been a key ally in the human pursuit of making our mark. Significant noise pollution during the pile-driving phase is somewhat guaranteed.

Remember – underwater, sound is sight. Ears are eyes. For porpoises, as with all toothed whales,[10] the density and opacity of water place acoustic communication at the heart of survival. Short, intense, high-frequency 'clicks' govern the language of their navigation, mate-choice, mother-calf bond, feeding success. These clicks are generated by forcing air through specialised organs called the 'phonic lips' in the nasal passages below the blowhole – radiating through a region of fatty tissue called the 'melon'. Really excellent.

When swimming, a porpoise clicks around 20 times per second, furiously increasing the number of clicks as it homes in on prey like a metal detector. If successful in a catch, the clicks end in a finale called the 'terminal buzz', at hundreds of clicks per second, that traps fish in an acoustic beam. Such feats of aural precision are observed across nature, in other toothed whales, of course – and bats. It is evolution in its element, and for the porpoise, the whole symphony sounds like the most outstanding fart. But receiving an 'echo' of their clicks from fish, an obstacle or another porpoise is key to making this behaviour worthwhile. My loves, there is no point in blowing clicks out your melon via your phonic lips if you cannot read the information upon return. It would be like sending a hopeful text to someone you fancy; they read it but never reply. The worst.

Anyway, the researchers in this German study wanted to measure whether the act of driving immense structures into the bed of the Danish North Sea, with pistons and

[10] Dolphins, killer whales, sperm whales, narwhals, etc.

hammers and such, might cause our harbour porpoises a bit of a headache. Well, shockingly, it did. But the long-term effects may not be as expected. Within an hour of pile driving, the colossal sound pressure generated underwater curtailed acoustic activity of the porpoises by 100 per cent: clicks, echoes – the lot. But I'm pretty sure humans would have had a similar response? All acoustic communication remained below typical levels, and general porpoise-related activity declined steadily during the five-month intense pile-driving phase. The researchers state that this adverse impact on porpoise communication lasted much longer than they thought. But they also reported a short-term *increase* in harbour porpoise chatter, at 22 kilometres *away* from the construction site. The impact might be far from temporary, though, as a 10-year study observing porpoises near the Nysted Offshore Wind Farm in the Baltic Sea found that acoustic activity among the porpoises here declined to a worrying level – and is still recovering.

The nature of the harbour porpoise's high frequency 'click' (one of the highest across toothed whales) means that to communicate, they need to do so within a space of one kilometre so that they can hear one another. Meaning they must have room to be *physically* close in order to communicate. As fellow mammals, we can surely understand? As my chat with Hanna from Swansea University intimated, assessing at the population level the long-term impacts of an animal as elusive as the porpoise is very difficult.

Nifty mitigation techniques like 'bubble curtains' have been tested around wind farms across the North Sea to try to shield marine wildlife from the noise. Bubble curtains might sound like the magical prelude to cake at a five-year-old's birthday party, but they are already proving relatively effective at diminishing the adverse effects of hearing loss and subsequent behavioural changes in our smallest cetacean. Immense volumes of bubbles are released via perforated hoses that encircle the foundation of a turbine. Connected to supercharges on ship decks, the bubbles serve to intercept and dissipate incoming sound waves. It's early days, and more research is always helpful, but it's getting more and more challenging to match science with the pace of climatic and societal change.

Hanna is an advocate of porpoise adaptability. We all should be. As I write, she referred to an 18-month study by her colleagues, published in July 2021. A complex hydrophone array measured echolocation around tidal energy generators, measuring consistent clicks close to the turbine. These promising data tell us that porpoises can evade the rotors of a turbine and minimise the risk of collision.

'Porpoises can and will adapt, but where their threshold is for this kind of disruption and subsequent relocation, we simply don't know. I think, quite strongly, that we must reduce our consumption drastically, so we don't have to produce so much energy in the first place,' Hanna implored. Kettles, lights, toilet flushes … human behaviour is hard (but not impossible) to change.

The harbour porpoise has a constant need to feed. I believe them to be my soul animal. Hanna also divulged that porpoises play. 'They will play for hours. Because having fun is important in life. They actually enjoy themselves, and that is quite fascinating.' Cousteau, too, marvels at the porpoises' games in his book *Silent World*. Chasing each other, swimming in spirals, blowing bubbles … sound familiar?

At Strumble Head, it's possible to set the clock to the feeding pattern of the porpoises. Put a turbine (or 50) there, and it will affect who comes to the dinner table – if any at all. 'But ten kilometres up the coast, it could be a different story. Porpoises might feed on the ebb instead of the flood – all because the fish in that area might behave differently,' Hanna pointed out. This is helpful to know if we are to test the compatibility of renewables with endangered cetaceans. She referred to a 'Survey, Deploy and Monitor' policy in Scotland, where renewable devices are installed under the proviso of a heavy amount of scientific research happening alongside. 'We need the devices *in* the water to find out about potential harm to cetaceans,' she added.

It is a curious case of fish. As the Mock Turtle jested with Alice in *Alice's Adventures in Wonderland* – 'No good fish goes anywhere without a porpoise.' Lewis Carroll, you tease. Harbour porpoises may be among the smallest mammals in the sea, but they have adapted their dense bodies to tolerate extremes of water temperature – from the eastern Mediterranean to the sea ice lapping Greenland. They are really quite legendary to have achieved such geographical spread. And how to manage a fairly constant risk of heat loss to chilly waters while on the prowl? Eat. Continuously. Day and night, they're on the binge. Their metabolism is nearly double that of a terrestrial carnivore of equal size. And when they're not eating, by God, they wish they were. A couple of marine biologists once perfectly described the harbour porpoise as leading a 'life in the fast lane' in a 1995 review of the life history of the porpoises in the Gulf of Maine. I love this. Our little harbour porpoise takes life by the horns – and sprints with it.

They have a broad taste for the pelagic and benthic (open water and seabed dwellers): herring, mackerel, sprat, cod, plaice and sand eels. Autopsies of stranded porpoises have also shown a liking for squid and, in younger ones, krill. It's a perfect example of how, for all predators, it's the groundwork laid by these lower members of the food chain that ensures a full, healthy stomach for the porpoise, the dolphin, the shark, the whale. There is no oak tree without acorns. No Sunday roast without pollinators. Less air for us to breathe without plankton.[11] A study in 2018 suggested that both captive and wild porpoises can alter their food intake and double their blubber thickness to adjust to the cooling waters between summer and winter. Blubber is a fatty layer of tissue beneath the skin of all marine mammals, providing a layer of insulation and energy – an inbuilt down jacket-plus-Christmas-belly sort of situation. Our own body performs a similar regulatory service via homeostasis. The general rule is that it's much better for everyone to maintain a stable body temperature than to waste time and energy flip-flopping between states – when you could be out eating, mating, fighting enemies and such. It gives porpoises innate freedom to move where they please, when they please.

Despite most of the cetaceans in the British Isles being used to fluctuating water temperatures, some scientists consider the possibility that our little harbour porpoise might one day, in a warmer world, suffer from *hyper*thermia – *i.e.* overheating. The (evil?) twin of the more well-known hypothermia. Data on the thermal tolerance of wild marine mammals is largely non-existent, mainly because they're so bloody hard to get hold of for any length of time to do this kind of research. We know that other larger mammals in polar waters, like bowhead whales, have dense, cushioned areas of blubber. But for species like the porpoise, it's fair to worry about how their physiology will manage these unforeseen extremes. They haven't planned for this.

Run with me here. We know a thick layer of blubber insulates against inevitable heat loss during winter. But the trouble is, the seasons are no longer fitting into the boxes we

[11] Plankton (tiny creatures drifting in the ocean) generate 50 per cent of the oxygen we breathe. They also provide a huge carbon sink. Pretty fundamental.

have made for them. They are spilling over the lines. There is cheating and hypocrisy – especially between winter and spring. It would only take an unusually mild winter for this neat layer of thickened blubber to be redundant. And our porpoise suddenly goes from being hyper-mobile and agile to a rather hot, buoyant little whale. It is mid-November as I write. Unseasonably mild; 15°C, wet and stormy. I wore shorts and flip-flops to the Londis this morning. Is winter losing its bite?

In time, heat stress may well drive harbour porpoises to up sticks and migrate north so that they can calve in cooler, more productive waters. But this gamble will only pay if that habitat is decent – *i.e.* plenty of food and clean waters with minimal risk of tangles with boats. In fact, it's already happening. The changing climate is pushing species between oceanic latitudes like chess pieces. Evidence of changes in cetacean composition around the British Isles due to rising sea temperatures is growing. Some Australian research published in *Nature* (2020) measured something called 'climate velocity' – the speed at which animals would need to move to remain within their ideal temperature range in the ocean. The findings revealed an ugly truth. Overturning previous understanding, these data predict that deep oceans could warm seven times faster by 2050 – even if we wildly reduced our greenhouse gas emissions.

No longer is the deep sea the safe, stable, untouchable abyss we like to believe. As I write, a fleet of volunteer vessels: ferries, cargo ships and containers – are trawling torpedo-shaped devices called a Continuous Plankton Recorder (CPR), as part of the planet's longest-running global marine survey. Running since 1931, this is an astonishing example of citizen science. Volunteers are dispatched in the name of data and planetary health to inform research. It's one of the only long-term datasets we have to make any kind of prediction on ocean health in the future. Focusing on plankton and its movements has decoded some of the impacts of climate change in the oceans, finding a general shift northwards of plankton in all areas. But these northerly areas are becoming a crowded waiting room of anxious ocean travellers. This is climate velocity in action. Greenhouse gases, the silent assassin rippling through all layers of the ocean at different speeds, is wreaking havoc on the precious connections and associations that exist between marine species – half of which we don't even know. It's a cruel cocktail. Bond would have hated it. Too much shaking and stirring. And not enough ice.

I was struck by the despair in Hanna's voice over the phone when I asked her if we would likely see jumps in seawater temperatures within our lifetime. She said we most definitely will. 'We're dealing with the feedback loops of previous loops,' she said. Greenland is key here. In many ways, the true thermometer of the planet. According to NASA and the European Space Agency, Greenland's ice sheet – the veritable celebrity in the frozen world – has lost 3.8 trillion tonnes of ice between 1992 and 2018. If, like me, you can feel your brain stumbling during your attempt to process these numbers, then picture this: 3.8 trillion tonnes of melted ice is the same as losing the water from 120 million Olympic swimming pools to the ocean – every year – for 26 years. Warmer air temperatures and fossil fuels are to blame for 50 per cent of this melting. Already, all this has raised the global bathwater by 11 millimetres (the width of my present brain) – and the rest of the ice sheet has a further 7.4 metres potentially up its sleeve. Gulp.

Aside from the fact that a further 6 million people will be at risk of flooding from around the world – the more sea surface exposed, the more carbon and other waste the ocean absorbs. This feedback loop can have cascading effects on all marine life in these waters. Weirdly, it seems the porpoise can temporarily monopolise the new feeding grounds that emerge as the ice melts. Scientists found an improvement in harbour porpoise body condition between 1995–2009, during a period of local warming in West Greenland. Less ice enlisted more cod and capelin larvae from East Greenland and Iceland – a symphony to a hungry harbour porpoise's ears. But please note – this isn't a reason to become complacent. We're seeing effects now that were forecast to happen between the years 2070–2100 of Arctic melting. Who needs a Tardis when our future is playing out before our very eyes?

I asked Hanna what she feared for the porpoise specifically. 'If you look at what they tend to die of, it's so often food-related,' she said. 'They cannot go long without food before they begin to starve.' If you've ever been out for a walk, run, or any type of outdoor frolic, you might be familiar with the feeling of a 'sugar low'. Those who are serious about cycling (stay calm) call this 'bonking'. I have had many an ill-timed bonk on my bike, and it's the most terrible sensation. Essentially, it describes a genuine physiological state when you haven't eaten enough carbohydrates and have depleted your body's glycogen stores, leaving you with that fizzy, shaky, sweaty feeling of having deficient blood glucose levels. These symptoms alone should be enough of a warning that an emergency absorption of five Mars bars is your only path to survival. An extreme bonk can even induce a coma, which would be a rather sour turn to the family bike ride. The point is, this is pretty much what happens to our harbour porpoises when they don't satisfy their incredible metabolic rate. If the areas they usually patrol are devoid of adequate fish supplies, they might very quickly descend into an irrecoverable bonk.

Let's pop back to Strumble Head with Holly for a minute. To the right and further down the coast, Holly pointed to aptly named Mackerel Point. 'Before any physiological impact, the biggest impact of climate change we're likely to see here with the porpoises will be to do with fish,' she mused. She told me how many of the locals in Fishguard and Goodwick recall when 'hundreds of mackerel … *hundreds*' were caught on a given fishing jaunt. But, as with insects on the windscreen, that was all in the past.

'When we do see issues with fish stock,' Hanna later explained over the phone, 'there can be a critical vulnerability if the porpoise cannot adapt quickly.' Some scientists dub the infamous see-saw of Greenland cod stocks through warm and cold periods as one of the most prominent examples of the climate acting on marine ecosystems. The International Council for the Exploration of the Sea (ICES) reported that British mackerel populations alone have been in a nosedive since 2011, dropping more than two million tonnes in weight between 2018–19 and losing their 'sustainable' status. The vulnerability of British fish stocks clearly must have had a serious impact on harbour porpoises – even though they are hardy little whales. Sigh. Hanna admitted the monumental task of isolating the specific threats that face the harbour porpoise. 'They can survive a lot, but we don't know how much all this affects their social processes, their family groups. We haven't got a clue as to *how* they actually live, you know? And I'm worried we'll leave it too late.'

By 3 p.m., my time as a porpoise spotter was rapidly waning. And then Holly said something which I have found impossible to forget. She said that we tend to only care about what we can physically see. I've heard this before – but it was the way she said it, in this extraordinary setting, that made me think. I thought of my family. My friends. The way a good red wine lingers on the glass. I thought of the blackbird that hops along our neighbour's roof every morning. My iPhone glinted in the sun. These things are all tangible, physical, within reach. So, what sickens me is that we are more likely to see a harbour porpoise emaciated in a fishing net than we are to see its dance with the waves. We know we have been keen on fish and fishing for nearly 200,000 years, but the very act of fishing, and the unregulated rate at which it's happening, is the immediate threat to porpoises in the British Isles – before the temperature change, before the fish depletion.

'Harbour porpoises get caught and die in nets,' said Hanna plainly, 'this is *absolutely* the top cause of mortality – from the Baltic, across the Channel – everywhere.' The World Wide Fund for Nature (WWF) and Sky Ocean Rescue found that an average of three harbour porpoises is killed every day in the UK as 'bycatch'. Bycatch describes the event of unwanted fish and marine animals caught accidentally during commercial fishing activity. Gillnets, in particular (nets that trap fish through their gills), tangle porpoises and suffocate them. As I write, 'sustainable fishing' seems no longer worthy of policy. Key 'sustainability' language has been written out of the latest Fisheries Bill (2020) by the Government. European waters are also coming under fire for failing to regularly monitor dolphins and porpoises, especially in bycatch. We're in danger of ring-fencing the most kinetic of environments with more red tape, around a sea that can barely continue to absorb our mistakes.

'My worst fear is that we don't have a government that wakes up to the ecological crisis,' she said. 'We're allowing mass industrial fishing to continue – and this will significantly disrupt the delicate biodiversity upon which porpoises depend.' Maybe it might be better to worry about something that hasn't happened rather than not worry and then receive an almighty surprise. Instead, she said, we should always expect to be harming something. That's how conservation works and has done so for years – abiding by a 'precautionary principle' – surmising consequence to every action.

One other thing that seems key is that overfishing and bycatch incidences release carbon into the atmosphere. A fascinating study in late 2020 revealed the enormous (and largely secret) carbon sink that we're losing – by not letting whales and fish sink to the ocean floor after death – preventing what's known as 'carcass deadfall'. Presumably, a natural process? Their data estimates that ocean fisheries have released nearly 1 billion metric tonnes of carbon dioxide since 1950 – preventing about 95 per cent of natural carbon in fish and whales from being buried. A large whale, like a humpback, can sequester around 33 tonnes of carbon once it dies naturally. Surely a no-brainer then for fisheries and governments to prioritise the avoidance of larger species? To prevent further temperature rises, we must save the whales.

Again, it's the issue of how we use this space. Because the trouble is, much of the area offered to UK offshore wind farms and renewables also clashes spectacularly with Special Areas of Conservation (SACs) for the harbour porpoise. Some research even goes so far as to propose that offshore wind farms could one day co-locate with Marine Protected Areas (MPAs), rendering the initial disturbance to wildlife negligible.

Plymouth University reviewed evidence in 2013 that explored the potential for the structure of a turbine to accommodate wildlife – whether the imposition of an offshore wind farm could exclude fishing activity, thus encouraging fish to school the area. Perhaps the footprint of a turbine could provide valuable refuge areas, allowing the accumulation of bivalves, crustaceans and other food web instigators?

But maybe we're trying too hard here. Certainly, Hanna had a word of caution: 'There simply aren't any easy solutions left – but we must consume less – less of everything.' Good point. For all its brilliance, technology is just part of the solution. Sure, a big, well-connected, jazzy piece of the puzzle – but just one piece. There is little point in dreaming up fanciful pictures of wildlife dancing happily about a turbine if consumption continues to rise. If we support fast fashion, tumble-dry laundry on a sunny day, leave the lights on and the engine running. Try as we might, we cannot just install some renewables and carry on as per. Not any more. Hanna insisted on the need for urgency. Urgency to reduce fossil fuels. Urgency to recognise nature as something with rights. 'The sea does not belong to us – we need to share it with many species – porpoise included. Some people use the term "other-than-human-beings", which kind of resonates with me.' And with me.

But not everything human-related spells natural disaster. On World Oceans Day 2021, the Department for Environment, Food and Rural Affairs (DEFRA) announced the designation of Highly Protected Marine Areas (HPMAs) across the UK in each regional sea, following relentless campaigning by organisations such as The Wildlife Trusts. HPMAs hope to act as top-tier buffers against the gaps in legislation around exploitation – by enforcing a total ban of all harmful activities. Such protection presents a fascinating experiment to see how nature recovers when stressors are reduced. And, we have officially entered the UN Decade of Ocean Science for Sustainable Development. Put simply, the G7 nations are clubbing together to drive positive change for oceans during the 'Ocean Decade' (2021–2030). We're making some good moves, but are we acting fast enough?

Back in my hostel room in Goodwick, I pondered my next move and fingered a shortbread from the tea tray. I figured I had two choices at this time of day: engage in a session of light, thoughtful drinking on the harbour wall, or hire a paddleboard and try and find these porpoises in one final attempt before I left the following morning. The latter, I felt, was more sensible given the waning daylight and, anyhow, Tesco Express and its beverages were conveniently on the way back to the hostel.

I cycled through Goodwick, up over the hill and down a steep one into Fishguard Harbour. I overheard a man on the phone describe Fishguard as the 'undiscovered Padstow, mate', which I thought very apt. The harbour was picture-perfect, the tide high and smooth as sea glass. I had booked on to the last place of an evening stand-up paddleboarding tour around the coast and was soon tagging awkwardly along with a family of four from Kent who definitely wished I wasn't there.

Our energetic instructor, Libby, was blonde, nose-ringed, steaming with coastal knowledge. She was everything I wished I could be in that moment. Fortunately, standing on a paddleboard is something I've become quite good at – it being my preferred mode of transport on the rivers at home to sneak in among reed beds to spy on coots and call to the beavers.

Compass jellies lit the water below our boards, dazzling like chandeliers in their turquoise ballroom. Pigeons stirred into flight as we meandered sea caves. The harbour porpoise had no doubt eluded me, and I felt a sting of disappointment at not seeing it that afternoon, but I decided I was idiotic for thinking that. It was nice to know they were there somewhere, and they would show themselves if they wanted to. They just have better things to do than to oblige me. I enjoyed myself immensely as the sun dipped below eye level, and we wobbled on our boards back towards the harbour, marinating in that rare preserve of a balmy July evening in Britain.

Thinking back to that moment, I recall what Hanna had said towards the end of our call. I had asked her what had drawn her to the sea and the porpoise. Referring to her time as a diving instructor and wildlife guide, she told me, 'Fish don't really care if you're there, but porpoises will look at you. You know that they know you're there. They give you *the look*.'

I had a greasy round of chips and rode back to the B&B as the sun set. It all felt very ideal. At breakfast the following morning, a chalkboard next to the doorway displayed a lesson from Proverbs 3:23 – a vague attempt to empower travellers up the steep staircase after a bellyful of beans on toast?

> *Then you will go on your way in safety, and your foot will not stumble.*

I crammed a third piece of toast into my mouth and fell up the first step on my way to get my things – time to leave. Snowdonia beckoned.

CHAPTER THREE

Seagrass

My first trip to Snowdonia was in 2003. A tepid diarist at nine years old, two entries reveal my journey of Welsh discovery:

> **8th August 2003 (just had dinner):** *Tomorrow we are going on holiday to Wales apparently! At first I was like 'oh great, ANOTHER boring hiking holiday!' But then I heard there are loads of sheep and things and now cannot wait. I'm a weirdo!!*
>
> **15th August 2003 (major problem):** *My feet hurt. Half of me wants to go home, half of me doesn't. Sheep were nice. Also, am becoming a weather geek.*

Sixteen years later, with a few Snowdon trips sprinkled in between, I was eager to return. There was just a slight issue – the train I was due to catch from Fishguard and Goodwick station to Porthmadog – the *11-hour* train – had been suddenly cancelled that morning. I wouldn't put it past Wales to halt all rail travel for a sheep that decided to nap on the tracks. Understandably, my immediate feeling was disappointment at not being able to relay the experience of passing through 45 stations along the iconic Cambrian Line. I wanted to sit by a smeary window like a lovesick poet. To bid a tidy '*hwyl fawr*!' to Wales, as we ventured into England to Shrewsbury, before crossing back through Snowdonia National Park, arriving, daring and whimsical, at the wild Welsh coast. I wanted to be one of the little floating heads in a trio of carriages, racing through a fleeting landscape. And I wanted to at least *attempt* to pronounce: Pontypool, Tonfanau, Dyffryn Ardudwy and Penrhyndeudraeth, without spitting all over myself as we pulled into each station. But, alas, it was not meant to be.

Perched on the edge of the bed in the Goodwick B&B, I instinctively rang my dad, bending back the corners of the now-useless train ticket as the phone rang. My remaining options were few. Someone was expecting me to be suited and booted on a beach somewhere the following morning to see what I had come all this way for. Time was precious, and at some point, I also had to return home to work. For all its splendour, Wales is not blessed with the public transport system of, say, Hong Kong[12] – and getting to Snowdonia without a car is tricky. Nor did I have time for, or fancied funding, a 100-mile hike or bike ride. A 500-hour bus journey at 2 miles per hour was not the vibe I envisaged for this trip.

But, thankfully, it turned out Dad was due to be driving up to Shrewsbury later in the week to visit my brother and his family. Convenient. And, unravelling himself from my little finger, he gallantly offered to travel a couple of days early via the 'scenic route'. He said if I could get myself up to Cardigan Bay by the afternoon, I could hitch a ride up the final stretch

[12] Famed for operating on one of the most efficient, reliable and sustainable public transport systems in the world. Following a work trip there in 2018, I can confirm it must be the eighth wonder of the world.

to Porthmadog. 'We'll make a trip of it!' he said. I could already hear him stuffing his sleeping bag into its sack. No, it wasn't the lowest carbon option, but it was the best I could do.

I don't remember much about the cycle to Cardigan; 25 miles of nauseating ascents and descents, following National Cycle Routes 47 and 82, across Pembrokeshire Coast National Park. Amid the cheer of a sunny morning, I soon relished the absence of train time and platform changes. The air was hot, and my eyes stung from sweat as I climbed the infamously steep A487 out of Fishguard Harbour. A roadside verge, bright green and wide, had been left to fend for itself. Grasses swayed in the breeze, and the absence of traffic allowed pollinators to steal the soundtrack. Pulling off the road and letting my bike fall, I lay down briefly on the grass, closing my eyes and catching my breath. White clouds skimmed the sky like fish darting to the shallows. Respite, indeed.

With a rucksack that felt several kilos heavier than the previous day, I felt encumbered. The back of my head repeatedly collided with hordes of snacks rammed in the lid. So, I favoured a downwards gaze, legs spinning furiously, my possessions constantly striving to throw me off balance. Some descents were so steep I had to dismount and walk, to save from toppling over the handlebars.

An idyllic village sandwiched a much-needed gap between the main road and another grossly steep hill. Shirtless and mowing a lawn near the church, a smiley, older man flagged me down and asked if I was quite alright before warning me to shift into my lowest gear – and best of luck with the rest of the ride. 'Argh! A *hefty* one, mind!' he chuckled, shaking his head and restarting the mower. I had come to recognise that a young, panting woman on a bike, alone and overshadowed by an enormous rucksack, for some reason gives the impression that something dire must have happened and surely she must need help.

A Land Rover overtook, its engine roaring as it met the incline. Through the tree tunnel, I pretended its disappearing engine was me and my bike, such power we generated together. But, once my front wheel started to lift off the ground with the sheer angle of it, my legs surrendered.

Let's digress – and take a deep breath. Ex*xxhale*. Now, repeat that twice. Breathing feels good, doesn't it? If anything, pretty priceless. Even better – two out of three of those breaths were enabled by the ocean, which is very noble of it. The staggering fact is that more than 50 per cent of the oxygen that humans, plants and animals breathe is born within the minute architecture of phytoplankton, the bygone solar factories that drift the braids of ocean currents. Absorbing carbon, water and sunlight en route, they convert it into food and release oxygen as a by-product. The life story that evolution wrote for these microscopic plants, algae and bacteria is the genesis of our planet. It is the ancient gift that enabled our ascension to the top. No coincidence, then, that photosynthesis is one of the first scientific processes taught at school, and flowers are among the first shapes we commit to paper. It makes sense, the very bottom of Earth's pile being a wise place to start. And, should we care to notice, the grassy meadows, forests and woody assemblages that shape our story are not restricted to a terrestrial existence, for they have underwater counterparts. Vast, untaught and unchartered.

These are marine vegetated habitats. Mangrove swamps, salt marshes, kelp forests and seagrass beds cover less than 0.2 per cent of the ocean surface area, yet harbour 50 per cent of the carbon buried in the global seabed. Seagrass beds alone account for at least 10 per cent of that global carbon storage. Marine vegetated habitats are the submarine fleet of the natural world. Our 'humble servants, beneath the sea'[13] authorising the web of chromatic, pelagic life – for which we lack time. I enjoy the irony, given that we – humankind – would be shrivelled, lifeless lumps without them. An arresting image.

Britain's story of seagrass in the last decade is as humble as the plants themselves. Co-founded by Dr Ben Jones (or @BoardshortsBen on Twitter) and Dr Richard 'RJ' Lilley in 2013 after their master's degrees, Project Seagrass is a marine conservation charity that sees the unseen and gives it a voice. Safe to say, their work in making us less ignorant of the sea and its secret garden is astonishing. Described as 'marine powerhouses' and repeatedly saluted as critical to global coastal marine ecosystems' health, meadows of underwater grass are not something to muck about with.

Nicknamed marine conservation's 'ugly duckling', what seagrass beds economise for in our imagination (picture a field of grass underwater – it is what it says on the tin) they redeem in captivation. Until I wrote this book, I was among the millions who were unaware of the submerged meadows that border the coastline of the British Isles, let alone even remotely awake to its importance and plight. Foxes were braver. Orcas felt bigger. Woodland, more tangible. But, the very idea of submerged meadows that annually flower and go to seed summer after summer, that have teams of crustacean pollinators and that house an entire world of dependent life, soon reeled me in hook, line and sinker. Some meadows are so vast that they could cover more than 400,000 rugby pitches. How these have passed us by for so long is a mystery. But I'm in the market for it.

Seagrasses are the only genuinely marine flowering plants in the UK. Seventy-two species of seagrass exist worldwide, on every continent except Antarctica. They

[13] From a version of the 1860 'Submariner's Prayer' – author unknown.

vary dramatically, both in size and rates of carbon burial. A researcher once beautifully compared Mediterranean seagrasses to oak trees, dwarfing the fronds of further north. Three *Zostera* species of seagrass dominate UK waters: common (often known as eelgrass), dwarf and narrow-leafed. Like its terrestrial cousin, seagrass favours a sheltered and stable environment, growing in shallow bays, absorbing sunlight and photosynthesising away just like any other plant. Veins within the leaf tissue act as highways, shuttling the sugar and oxygen produced by photosynthesis around the plant. Pockets of air within those veins act as tiny buoyancy-aids – lifting the leaves to the water and keeping them from sinking.

Seagrass is both the native and the outlander. Marginalised to the extreme. Just how many seagrass beds flank the UK coastline remains unclear. But, what we do know is that the meadows that linger are regarded by scientists as in poor condition. At times, they've ranked among the worst in the world. Tissue nitrogen levels have been at least 75 per cent higher than the global average. As a general rule, excess nitrogen in plants is not good news. Stunted root growth, dehydration, and surrounding water pollution are just some of the stressors that a given plant will endure in these circumstances. I think it's fair to admit those people working to protect seagrass have cause for a monster headache.

But good work is being done – and fast. One of the continual challenges in conservation is generating awareness across a broad audience, hoping to inform good decisions. How can we expect people to muck in and help if they don't know the score? If they don't know what or why they're helping? Apps like SeagrassSpotter encourage and enlist public support in mapping locations of seagrass beds throughout the British Isles – so limited is the knowledge of their modern-day whereabouts. There have even been studies investigating the role of psychology in finding ways to defeat the problem of seagrass's repressed public image! I mean, come *on* – seriously? Desperation is growing – and with good reason.

The decline of this habitat is of mounting international concern. The latest figures looking at historic maps suggest the global oceans could be mourning the loss of 92 per cent of their seagrass. Disease, extreme weather and physical disturbance from us are the main offenders – but we'll talk about that later. On its website, Project Seagrass equates this loss to the size of two football fields disappearing every hour – since 1980. And we continue to lose an estimated 7 per cent of seagrass around the world every year.

Restoring seagrass meadows worldwide alone will contribute to reaching 10 of the United Nations' Sustainable Development Goals. In its classic 1994 paper (you'll hear my gasp from the Congo if you've actually read this), the UK Biodiversity Action Plan (BAP) identified seagrass beds as a habitat to prioritise. This is surely the least it could do for a habitat that absorbs and buries carbon up to 40 times faster than a pristine tropical rainforest. This oft-quoted 'blue carbon' is the underdog we need to seriously start cheering for, as based on that estimate, the UK would need 70 hectares of tropical rainforest to match this unlikely carbon-sapping skill. A situation where one might suggest that *you do the maths*.

Conservationists are increasingly recognising a tendency for their practice to take a 'speciesist' approach, using the celebrity of puffins, for example, to boost public intrigue. An attitude that focuses on saving one species at a time. To a certain extent, this book aims to do precisely that. Although of value, this method runs the risk of cleaving a living thing from its very surroundings – of failing to acknowledge the thread that forms the tapestry. Without habitat, there are no species to market, no flagship individuals for which to gain membership. Unfortunately, we also don't have the luxury of time to think this way any more. The resulting landscape is soon pocked, scarred by the life it failed to parent.

I once noticed a farmer on Twitter who was carbon-accounting. He expressed surprise to learn that his grassland was absorbing twice the carbon (per hectare) of the nearby woodland. How refreshing to witness his thinking revive, as he called for his followers to see woodland and grassland as a rich, complex patchwork, not some standalone silver bullet. After all, the soil below terrestrial grasslands harbours some of the highest carbon stocks of any UK habitat. The same goes for seagrass.

And so, I wanted to find out why the fields below the seas were suddenly being woven into the British climate narrative with such urgency. Porthdinllaen, a tiny coastal village embraced in a bay on Snowdonia's spectacular Llŷn Peninsula, remains home to one of the most extensive seagrass beds in the UK. Time to see what all the fuss was about.

Arriving in late afternoon, I regretted not having more time to explore Cardigan Bay. Famed for its marine wildlife, this is the first area of coastline in Great Britain to be designated a Marine Heritage Coast and now another Special Area of Conservation (SAC) too. One of the largest bays in the British Isles, its 60-mile expanse offers one of those moments where suddenly mainland Britain materialises on the map in front of your eyes.

Sandwiched in the middle of the supercontinent Pangea (around 250 million years ago), Britain and Ireland enjoyed closer company. Over time, they continued to separate as Earth's enormous jigsaw began to disassemble. Rejoining briefly 10,000 years ago during the last Ice Age via the aptly named 'British–Irish Ice Sheet', they shared similar geology and biodiversity. Many of the notches, edges and grooves that today characterise both coastlines owe their anatomy to this ice and the slow retreat of its vast glaciers. If you see Cardigan Bay as half of a broken heart on a map, then its other half waits across the Irish Sea somewhere near Wexford. The arc sweeps from the Llŷn Peninsula to St David's Head in south Pembrokeshire, facing the prevailing Atlantic wind. Bottlenose dolphins, harbour porpoises, grey seals, razorbills, guillemots, gannets, gulls, sea and river lampreys, shags, basking sharks, leatherback turtles and sunfish call this shallow bay home. Either way, Cardigan Bay is a credit to Wales and the entire British Isles.

I was still unsure what I expected the town of Cardigan to be like, but it was different. A hub of Welsh heritage, culture and easy access to exquisite landscapes make this pretty town an easy sell. After failing to identify a single person wearing a cardigan, I loitered outside a pub that overlooked a brown Teifi Estuary. I spied a table that had a view of it and tried to blend with the youths of Cardigan in the sunny pub garden. Dad phoned, saying he was about 45 minutes away. Translation: 45 minutes to get up to all sorts before parental supervision prevailed. So, I ordered a pint of Guinness – my first since the pubs re-opened in the summer after lockdown. Sipping almost sceptically and spilling a decent lug on my shorts, I decided I was OK with feeling a bit of a novice on my journey. Drinking alone at a pub whose name I couldn't pronounce and a hopeless flirt with my bike, I ran a hand through my mess of tangled hair and massaged my aching quads. A strange old time.

Oh, wonderful, beautiful, comfortable car. A sight for sore bodies indeed. My earlier guilt at continuing my Welsh travels via diesel engine diminished at the sudden exhaustion and headache that hit as soon as I heaped myself into the passenger seat. *I'll make up for it tomorrow*, I thought. To the background of a classic 1980s playlist, chatter and comfortable silence, it didn't take long to slide back into childhood. The same turns of phrase preserving the legacy of my relationship with Dad. Father and daughter. Daddy and Sophie. On the road again. New times and old times. We wound north on empty, wide roads that mapped the land I had grown to love. Sunset stained the horizon over a slate-grey sea. Cottages lit the darkening valleys like tea lights, and I dozed most of the way to Porthmadog.

We awoke to biblical rain lashing the window of the Travelodge. Naturally, I stood by the window in my pants, filmed it and have never watched it since. There is nothing quite like receiving North Wales at face value. This was the Wales of my childhood – sunny days a bonus. I knew it well. My legs ached almost audibly from the ride the previous day, and I lingered in bed.

If locals call the Llŷn Peninsula 'Snowdon's arm', then my destination was around about the elbow region – a sodden 18-mile ride away. Keen to explore the Peninsula with me, Dad became my support car – taking my great lump of a rucksack and leaving me blissfully unencumbered and exposed to the rain. If you've never been lucky enough to visit this Area of Outstanding Natural Beauty (AONB), then put it on your list immediately. Imagine cherry-picking all the best ports, fishing towns, farmsteads, bays, cliffs and coves of Cornwall and confining them to a 30-mile emerald spit surrounded by the Irish Sea. A final sweep of lux, vignette and saturation (for you, influencers), ensuring the colours leap in all weathers, a smattering of lapping waves and just enough gull cries – and behold, you have your Llŷn Peninsula. And, dare I say it … Cornwall's more attractive twin. (Dramatic pause.)

I was happy to leave Porthmadog, wending through a dreary industrial estate and following all signs pointing to where fewer people would be. The geography changed. Shy mountains near the Snowdon range rose in the distance, their summits veiled behind clouds. The hum of an awakening Porthmadog had long since faded, replaced by oystercatcher whistles, a skylark and slick tyres on wet roads. Feeling thoroughly rinsed, I kept moving, resisting a toilet spot that had a vista of a bottling grey seal. Being overlooked by a fellow mammal during such delicate moments is something that even I cannot abide. Starlings lining webs of telephone wires welcomed me to Morfa Nefyn. Its rich local maritime history features Viking warriors besieging Norman castles on the Llŷn and in nearby Anglesey. One mile away in Nefyn witnessed the triumph of the Edwardian conquest over Wales in 1284. Sometimes it's the most unassuming of places that wallow in a reservoir of legend. Separated by a short headland, the crescent beach of Morfa Nefyn resembles a curvy 'W': one half being Porth Nefyn, the other Porthdinllaen – which is where our king-sized bed of seagrass lies. I skidded around the hairpin bend that steered me along to Caffi Porthdinllaen, where I bumped into Dad and was due to meet up with Jake Davies – marine biologist and master seagrass diver, who was going to take me on a bit of a snorkel safari.

It never occurred to me that Jake would not be a balding, middle-aged man. We had only ever communicated via email, you see. Dad and I were surprised, then, when a strapping 25-year old strode across the car park towards us. He had this Bond-like/Navy Seal vibe going on, dressed all in black – making a slight detour to help a girl out before his very daring and dangerous underwater mission. Snorkelling gear was hoisted gamely over his (must admit, broad) shoulders – a shark tooth necklace around his neck. Welsh people are, on the whole, some of the friendliest people you will meet. Wasting no time in getting straight to the point, Jake led us along the quiet, stony beach towards

Porthdinllaen, chatting about his work and upbringing in this picturesque corner of Britain.

The Welsh are a patriotic bunch, and it's lovely to see. Seventy-three per cent of the Llŷn Peninsula population favour fluent Welsh, and speaking to Jake, I sensed genuine pride in his community. A rare thing when most 20-somethings still seek a sense of place far away from home. He spoke of his father, who comes from a long line of maritime folk like so many here. The sea and its contents were woven more seamlessly into human life than today. He reminisced about their fishing at sunrise together before school. The very curve of this bay is a lifelong frame of reference for Jake's view of the world.

He explained how often Wales is overlooked. Whether for wildlife, research, long-term residence or economic promise, it's not always first to spring to mind. Following a degree in marine biology at nearby Bangor University, Jake works as the project coordinator for Angel Shark Project: Wales, led by the Zoological Society of London (ZSL) and Natural Resources Wales. It aims to defend and understand this critically endangered shark that so few have previously heard of (me included). With Wales being the northern limit for several key species in the British Isles, Jake explained how this place is potentially one of the last refuges in the world for the angel shark, outside a stronghold in the Canary Islands. Jake has also worked closely with Project Seagrass over the years, helping it carry out surveys, with seed collection and as an ambassador to raise people's awareness of the importance of these unique meadows. Like a young Jacques Cousteau, Jake's photography is an essential part of his arsenal in sharing these stories.

Fostered by the National Trust of Wales since 1994, Porthdinllaen too shares a fascinating history with the rest of the peninsula. The remains of the Trwyn Dinllaen Hill Fort on the western edge of the headland dates back to around 100BC. Visitors may relish the thought of this idyllic harbour once being considered a significant port between Dublin and London in the early eighteenth century before its revamp into a shipbuilding, fishing, trading hub some years later. Today, a small number of buildings hug the base of the headland at Porthdinllaen: some white, some red, others traditional stone. Some were so short and worn that I felt I might be both upstairs and downstairs at the same time if I went inside. A steely Irish Sea at low tide lapped the shoreline, undressing ladders of rockpools and seaweed. A green meadow lay somewhere below, just out of sight.

There's a tropical mood to seagrass that I cannot shake. Evie Furness, a marine biologist and technician for Project Seagrass, agrees. 'Most of the images or clips we have seen from seagrass tend to be from the tropics, where turtles and manatees feed,' she said. 'But who knew that we have it here?' As was becoming the norm, we spoke over Zoom, the conversation easy and bright, interrupted by the occasional delivery at the door. Evie is in her 20s too, and she and I share a similar journey. After school, we were both clueless as to what 'the plan' was and pursued broad undergraduate degrees that cherished interest versus expectation. A chance dive in a quarry near Heathrow Airport was enough to convince Evie that there was more to life than what exists on its surface. Hooked, she studied marine biology at Swansea University, interrupting travels to do a year in industry with Project Seagrass, before later accepting a position to be a part of its pioneering restoration mission. As I write, she's currently juggling work and a master's degree, and I admire her immensely.

Unlike much of the conservation narrative, the story of seagrass is uplifting and bewildering. It makes you dare to believe in a hopeful future for the aching seas of the British

Isles and to trust in the people fighting to save them. Seagrass Ocean Rescue is a one-of-a-kind initiative. Its name, though dramatic, is an understatement, for this is a project like no other. Seeking partnership with Sky Ocean Rescue, WWF and academic weight from Cardiff University, Swansea University and local support – the UK's debut large-scale seagrass restoration project is a celebration of collaboration.

It follows the footsteps of the iconic seagrass revival in the Chesapeake Bay – the largest estuary in the USA. A huge citizen science gardening endeavour ensued in 1999, where teams of researchers and the public rallied to plant 72 million seeds. The subsequent return of 9,000 acres (3,642 hectares) of seagrass to the Atlantic remains the single largest area of restored seagrass in the world. This achievement and Seagrass Ocean Rescue confirm how working in concert can discharge a dying habitat from intensive care. It's an approach that we must deploy far and wide. Our weapon of mass construction.

'I think a big thing is just how many people don't realise seagrass is there,' Evie admitted. 'We've lost ninety-two per cent in the UK over the last hundred years. It's insane how *quickly* we're destroying our coastlines without really having explored them.' I want to stress how the surge to revive seagrass is not just another PR spin for conservationists. Its marked disappearance has put our decaying relationship with the ocean under increasing scrutiny, despite large-scale declines of these meadows being a vague storyline throughout the last century.

The 1930s witnessed seagrass beds on both sides of the Atlantic hit with an epidemic of 'wasting disease'. Evie explained that this has a similar deleterious effect to potato blight, which triggered the infamous nineteenth-century Irish Potato Famine – the Great Hunger. Wasting disease led vast meadows to succumb to a pathogenic slime mould. Green leaves choked into black and brown – erasing 90 per cent of eelgrasses from North American and European seabeds. Combined with the barrage of artificial fertilisers, pesticides and herbicides that defined the intensification of agriculture, seagrass beds stood little chance of resuming their former glory. With these excess nutrients, sewage and waste leaching their way into the sea, fuelling algal growth and choking the seagrass's ability to photosynthesise – humans were rapidly (and unwittingly) robbing British coastlines of their shoreline identity.

A study on seagrass in the Thames Estuary (2018) found that water pollution from herbicides increased this meadow's vulnerability to future bouts of pathogenic disease. 'With so much pollution entering rivers and running off the land,' Evie insisted, 'and when you've already got a weakened ecosystem that is then hit by all this pollution, it really is the last straw.'

It won't surprise you to know that physical disturbance is also a massive issue for seagrass. We regularly prove how grass and propellers are bad neighbours. The ecology of perfection has preordained a trade-off, where steel blades triumph over cellulose ones. Much research has shown how disturbances in harbours with seagrass have lowered shoot density and bruised overall productivity within the meadow. The trouble is that both people and seagrass gravitate towards the same sheltered, shallow habitats – causing a regular stand-off. 'Unregulated leisure craft is a problem,' said Evie, 'people gunning it on a speedboat in a sheltered harbour on a sunny day, not realising seagrass is below, can result in the propeller acting like a lawnmower – cutting up the grass.'

The Seagrass Ecosystem Research Group from Swansea University, led by Professor Richard Unsworth (Evie's boss), estimated that at least 6 hectares of total seagrass loss in the UK is attributed to 'swinging chain boat moorings'. Dredging and anchoring, too, rip literal holes in the meadow. Reading extracts of Cousteau's *Silent World,* it's hard to ignore his genuine bewilderment – and horror – upon witnessing the damage anchors and trawlers can wreak on the seabed. He writes how fish escaped in terror, as though the net was the Grim Reaper, how nature's delicate farmland was swiftly obliterated and undermined.

If that wasn't terrifying enough, groundbreaking research (2021) on *Posidonia oceanica* seagrass in the Mediterranean – the oldest species on Earth – shows its sensitivity to human-made noise in the ocean. Noise created from boat propellers, for example, caused the loss and distortion of the seagrass's starch grains – stripping our vital plant of its crucial energy store. Growth is impaired and confused. Being opportunistic pathfinders, water currents leap on the chance to charter new routes through a vulnerable seabed, increasing the entire habitat's likelihood of even further erosion. All current research points to an urgent re-evaluation of seagrass-friendly mooring systems as we continue to enjoy the coastal playground. It's one giant shambles of a feedback loop.

Back on Porthdinllaen, Jake and I suited up. Despite the season, it felt chilly, and I envied the thick, fleecy lining of Jake's diving suit, hood and gloves – laughing weakly at my summer surfing wetsuit with its bright pink arms and slight Eau-de-Estuary. Not exactly the Lara Croft transformation I had anticipated. I looked and felt a bit of a thumb. No matter, though, as Jake had lent me a mask, snorkel and fins, and all I cared about was the clarity of the underwater field. As well as being my guide, Jake had a task to execute for Swansea University. At this time of year, the grass was nearing the end of its growing season and preparing to flower and go to seed during August, and

Jake was collecting shoots with seeds in a little pouch. This vital fieldwork has facilitated Seagrass Ocean Rescue in its revival mission further down the coast. We waded out towards the lines of little fishing boats anchored around the harbour, floating as soon as possible.

Looking back, I think swimming among seagrass could rank in the top five of the great human experiences. We forget the tender joy the natural world can illicit. Its raw force muffled under a duvet of digital distraction. The beauty of searching for a plant is it involves no spectacular chase. Rarely does it require hours of boredom in a hide and a rediscovery of how fascinating one's fingernails are. Plants don't try to outwit you. They are just there.

I recently asked a botanist friend of mine what gets him going about plants – what really *does it* for him? He told me plants face all the same stressors as other species – temperature extremes, food availability, space, light, predation, noise pollution – yet they cannot run away, fly away, hide or hunt. 'They have to evolve ways to deal with all these things while remaining rooted to the spot.' Pretty impressive. I had never thought of it like that before. So, I put my head under the water at the earliest opportunity. It was cold, clean and unbelievably clear, thanks to the natural filtration of bacteria, pathogens and pollution passing through every leaf. Gliding above a seabed that ripened from the rocky, sandy shore, the meadow unfolded before us like a concertina.

I felt as though I were entering the gates of Eden itself at times. This was a world that felt new and ancient all at once. As though time had no place here. Finally achieving flight, we soared. The heavy cloud that kissed the horizon didn't matter, as it felt as though light persisted within these blades of grass, causing them to glow faintly. Perhaps residual sunlight had entered the previous day and remained safely guarded by a million tiny cells. I still imagine this. Although undeniably similar, the leaves are longer and broader than the grass of our weekend picnics. The ocean frontier demands changed anatomy, urging blades in established meadows to grow up to 1 metre in length.

A freshwater eel darted into a burrow of blades. A rare sight for us, perhaps – but not for seagrass. Gliding weightlessly over the seabed, Jake pointed at it to make sure I had seen it, later telling me that it was only the second one he had ever seen, despite years of diving here. I duck-dived to join, my head caught in a vice of freezing Irish Sea – but I didn't care. My hand dangled, fingertips brushing the grass tips. Some blades were visibly longer. Others were only just beginning.

Yes, a seagrass meadow carries a simple aesthetic, but it has a three-dimensional quality that makes it so much more than just a field underwater. Denser areas of grass revealed the direction of the current, as a field of barley does the wind. I soon realised it wasn't just grass – it was a complex habitat in constant motion, governing the species that depend on it with immense grace and fluidity. I could only admire. Because if one were to compare our movements in the water – Jake waltzed where I dad-danced. Although lucky enough to have spent a good portion of childhood in/on/under the sea, my legs pulled me back to the surface every time I dove, deployed as though airbags. It felt good to finally realise my mermaid dreams, in North Wales of all places.

I want you to know that seagrass reproduces *sexually*. There is no other way of putting it, I'm afraid. Like terrestrial grasses and flowers (the lily is seagrass's closest land relative), male seagrass flowers release pollen from the stamen into the water column. Not only are they waterproof, but seagrass pollen grains are also the longest on the planet, measuring a whopping 5 millimetres, compared with the average 0.1 millimetres of other plants. Clumping together in horny, purposeful chains, they drift the currents. They seek out lady flower parts to delicately fertilise with their hefty member of pollen, sometimes for miles.

Later in the summer, crabs, fish and shrimp would likely assume the bee's role, moving from hairlike flower to hairlike flower across the meadow – valued assistants in this primal act. A group of researchers in 2012 looking at seagrass pollination in the Caribbean found for the first time that many crustaceans had seagrass pollen embedded in parts of their bodies. Coincidence? I think not.

There was a clear method and something rather hunter-gatherer-y about the seagrass sampling technique. Simple and satisfying. As though when finally muted underwater, we shrug off our cloaks and dress more appropriately, on a par with the seabed. Jake scanned the meadow for young shoots with seeds, gently picking, checking and gathering into a black netted pouch. The whole charade was not dissimilar to blackberrying along a September hedgerow. Taking a couple of shoots, I couriered them to the surface for further inspection. Picture young peas in a pod, and you're not far off. Although the seeds were visible, the leaves on either side of them enveloped them protectively, but they looked surprisingly fragile now loosened from their stable tether, their shelf life all too easy to see. I felt like an excitable child clutching a magical green wand, chosen from its underwater wizarding world for a covert assignment – the dutiful wonder plant, the lungs of the ocean, held in my frozen hands. I don't think I could have felt happier.

During more than 300 hours of dive time around the UK coast, millions of seeds have been gathered like this, fuelling the restoration goals of Project Seagrass. When I spoke to Evie, it was early December 2020, and she was both elated and exhausted. They had achieved their mammoth goal and returned one million pine-nut-sized seeds to the seabed the week before. 'We're in the recovery stage now,' she laughed. A phenomenal feat in an extraordinary year.

It occurs to me that humans can be genuinely remarkable beings. Somewhere deep within our sinews and synapses exists the will to make things happen. It's quite something. Remarkably, those involved in such feats seem empowered, buoyed – inspired, even – when they realise a goal. Maybe we should do this more?

As with most conservation endeavours, scale was crucial. The more seagrass planted, the better its chances of survival. Two hectares – the size of two rugby pitches – is the ultimate goal in Seagrass Ocean Rescue. A daredevil wish to establish a relict meadow for up to 200 million invertebrates, rare seahorses, pipefish, cuttlefish, cetaceans and perhaps even our endangered angel shark. A wish to armour our coastlines against warming temperatures, high carbon concentrations and the assaults of extreme storms and industrial fishing. South of where I snorkelled with Jake, the small community of Dale lies on the Marloes Peninsula in west Pembrokeshire. Evie explained how evidence suggests a carpet of seagrass once covered the entire Dale Bay, yet modern times have seen it regress into an anoxic, algal-dominated state. Ho hum, couldn't we just cover it up with AstroTurf, found a Boules Society and apologise, whilst pouring ourselves a Pimm's?

With research rapidly confirming seagrass as a serious ally in our fight against climate change, a heroic intervention was called for. It was a case of collecting enough seeds and planting them in a way that wouldn't damage the seabed but would give the grass a decent chance of germinating. All seeds were painstakingly handpicked, just like Jake was doing on the Llŷn. Evie reminisced the seed-collection dives fondly, 'Dolphins swam with us, sharks, seals, rays, fish – a total biodiversity hub.' Please, take us there.

All seeds were then sent to the lab at Swansea University, separated from their leaves. In theory, when seagrass rots, the leaves float to the surface, and the seeds sink to the bottom. For Evie, it turns out a little more coaxing is needed to separate seed from parent, the majority having to be sorted in the lab. 'We were up to our elbows in rotting seagrass for months! The lab smelt like a giant, rotting egg.' Worth it, though.

Once seeds were harvested, it was time to sow. In the early stages of the project in 2015, seabags (importantly, not 'teabags') were trialled in the Helford River, Cornwall, and local Porthdinllaen. Richard Unsworth led the team that included Evie and our friend Dr Hanna Nuuttila (seagrass, porpoises, it's all connected). A simple method referred to as 'Bags of Seagrass Seeds Line' (aptly nicknamed BoSSLine) planted seeds into the seabed within biodegradable hessian bags, released at 1-metre intervals underwater via lengths of rope. Such trials highlighted the importance of selecting a suitable environment. Remember, seagrass thrives in shallow, sheltered and stable bays. Ninety-four per cent of bags placed in such environments yielded successful shoots, compared with a location on the Llŷn that ruined the young shoots following a storm. 'We've got such powerful tides here in the UK,' admitted Evie, 'that if you just chuck seeds out in any old bay, they'll get washed away and eaten by shore crabs.'

With 20,000 hessian bags, it was time to rally the troops. According to Evie, an army of 2,000 volunteers included hundreds of local schoolchildren – 'The *best* workers you will ever come across. You're giving them a chance to play, and save the world.' Seagrass Ocean Rescue sounded like an enormous disaster relief effort, and I want to know where

I can enlist. A stirring thought that warrants a trumpet or two and some above-average wartime propaganda – I'm shouting 'DIG FOR BRITAIN, LADS!' as I write.

Each hessian bag needed to be filled with sand first, then topped with 50 seeds. Holes in the bags facilitate seed germination and offer effective protection from passing crabs and currents. Finally, every bag was tied onto 20 kilometres of rope – forming the biodegradable BoSSLine. Evie described how volunteers, clearly enamoured at the prospect of being part of such a movement, came from all over the UK in February 2020 'to lift heavy boxes of sand and seagrass and just plant tiny seeds in the sea – it was incredible.' Google the project, and the photos of these exploits are properly uplifting – humans at their best.

By the end of the week, the team had planted 750,000 seeds. By the following winter, they surpassed 800,000 – a small, restricted team planted the remaining 250,000 shortly before Evie and I met. The whole procedure is deliciously low-tech and replicable. Fifty volunteers were put up in a run-down fort in Dale for a week, up at the crack of dawn, prepping seed bags till sunset. 'Then we had the boat team,' Evie told me, 'lowering the bags tied to the rope onto the seabed.' Despite the Covid-19 pandemic affecting the project, most of the planting had thankfully been completed before the effects took hold. The team returned during the summer to see if seedlings were taking root and make any necessary repairs to the sediment. Evie recalled the lovable first shoots spiralling out of gaps in the hessian, which one day promise to flourish beyond all expectation. Monitoring the seabed around the meadow like this also keeps tabs on its efficiency at storing carbon. A 2015 study in the Marine Pollution Bulletin found a 20 per cent reduction in carbon stocks on the sparser margins of a meadow compared with the centre.

It's still early days to determine the long-term success of the seagrass revival in Dale, as is the case with any inaugural effort. There is no guarantee that the meadows will fully germinate. Still, the methodology is sound, thanks to years of trials and improved understanding from the teams at Swansea and Cardiff University. So degraded is the water quality that UK coastlines are currently unable to institute a 'natural' revival, meaning that further human intervention is needed. Next, the plan is to resurrect the estuaries within our industrial nerve centre: the Orwell, Humber and Stour, and additional sites in North Wales. Although magnificent, armies of volunteers might become an anecdote. The project in the Chesapeake Bay implemented combine-harvester-type equipment to collect and sow the seeds in favourable locations along the seabed. Mechanisation is an unlikely comrade in accelerating the future of this habitat in the British Isles.

Seagrass meadows support such an eclectic community of life that they have often been compared to an African savannah. Just as the acacia tree eventually feeds the vulture, flows of energy and matter swim between trophic levels, linking and reinforcing the coastal food chain, enabling life, death and rebirth all at once. In your average seagrass meadow, sea urchins and shellfish, some shy, others bold, along with invertebrate grazers, scavengers, decomposers, predators and prey congregate amid the fronds. Many of the species within a seagrass meadow, like oysters, have strong calcareous shells – an exoskeleton mostly made of calcium carbonate and a small amount of protein. Your average seashell, as well as corals and other molluscs, is dependent on this mineral housing for its survival – and scientists are thinking that the presence of seagrass helps fortify both the shells and their delicate inhabitants against looming ocean acidification.

The reality is that our oceans strongly feel an increase in atmospheric carbon produced by fossil fuels. So much so that too much of it can serve to lower the pH of the water. In water, carbon dioxide quickly becomes carbonic acid, soon disrupting the careful balance of acid/alkaline. Perhaps you've seen photos of statues or sculptures eroded by acid rain? The Taj Mahal is one of them – its glorious white facade is turning peaky and sour. Ocean acidification is the same concept. A relatively new area of ocean research, it is fast becoming a priority to understand. For animals like oysters, coral, and many molluscs with calcareous shells, a lower pH causes their shells to dissolve effectively in real-time. Already, figures estimate the global ocean pH has lowered by about 30 per cent. I imagine gradual disintegration would be a very unpleasant experience, no doubt leaving the victim feeling exposed and vulnerable. Animals that produce these mineral structures can repair damage and thicken them to an extent, but the effort of sustaining such maintenance would no doubt outweigh any long-term benefit.

But here's another reason why seagrass is just way above par. As you'll read in just a minute, seagrass can actively *remove* carbon from the atmosphere. And by doing so, it can rebalance the scales – and neutralise an acidic pH. Just by being a plant, seagrass can fortify the shells, the meadow, the *planet* against the threats we are creating. With experts suggesting that ocean acidity could increase by 150 per cent by 2100 if we carry on burning fossil fuels, whether or not we act on seagrass-related services should not even be a discussion.

With this in mind, it would be unwise to see the new meadow in Dale Bay as anything but an extraordinary example that should be followed. In 2009, the National Oceanic and Atmospheric Administration (NOAA) recorded the highest global level of carbon dioxide wafting around our atmosphere for more than 800,000 years. A concentration nearly 40 per cent higher than before the Industrial Revolution kicked off in the late eighteenth century. Global emissions need to fall by 85 per cent by 2050 if we are to prevent Earth's thermostat from rising by more than 2°C.

Essentially, we need to suck carbon *out* of the atmosphere – and quickly. Feel free to disagree, but seagrass, mangroves and marshes far surpass ancient woodland and tropical rainforests in this regard. They are in a different league altogether, I'm afraid (secretly delighted). 'What's amazing,' Evie admitted, 'is how much campaigning there is to save tropical rainforests – which of course are still vital – and yet we've got our own equivalent here that is consistently overlooked in its value to us.'

Seagrass beds offer stability and lots of it. Mental stability, of course, for a freezing face in the Irish Sea, but it's the architecture of a mat of roots that should prick the ears of the savvy. A dense subterranean rhizome fortress physically strengthens the seabed. Rhizomes describe an underground stem that in some plants grows horizontally. As it matures, a thick mat of new roots and shoots develops – lots of carbon in here! As marram grass reassures a shifting dune, seagrass meadows are a buffer, offering vital friction between sea and land.

The scenario goes like this: a storm brews out at sea, waves grow in strength, and they begin sprinting to the coastline. But, oh! Whatever is *that*?! An unforeseen obstacle has caused a sudden loss of energy in our lovely waves, so close to land. As though meeting an enormous speed bump, the wave's energy immediately weakens against the meadow below the surface. The seabed has robbed the storm of its power. Despite this inconvenience, the wave eventually does reach the coast, albeit with a muffled impact and a bruised ego. Speaking for all 60[14] seaside piers across the British Isles, make no mistake – we *want* this to happen.

Almost always, the structural complexity of a seagrass bed (with its roots, rhizomes, and svelte leaves) keeps it open to receiving carbon via both photosynthesis *and* external sources. Phytoplankton, fish turds, seabird turds – anything organic drifting in the water column will do. Leaf turnover is high, with older leaves falling to the seabed. Not dissimilar to a leaf carpet in autumn, yet where a beech leaf may rot within a few weeks, seagrass leaves will linger in low oxygen levels, stockpiling carbon the entire time. If you are in any doubt, wander an eelgrass bed at low tide (I've heard this is particularly rewarding on the Swale, near the Thames Estuary). You can reveal its dense, black carbon larder simply by scraping away the top layer of sediment with your boot. Together with their large surface area, seagrass beds can match-fund a tropical rainforest's centuries-long carbon donation, and then some because our talented carbon grave-digger can store carbon for millennia. It already has done. In the UK alone, our seagrass beds once stored nearly 12 million tonnes of carbon – that's the annual emissions of around 8 million cars. Yes, seagrass is the one that almost got away, but this reminds us of its potential to return and realise its former glory. As faithful as a farm cat (when given a chance).

Scientists in Spain's Portlligat Bay once found carbon deposits in seagrass beds measuring more than 10 metres thick, at 6,000 years old. Such is the alchemy of many plants – time travel is possible. Just like the rings of an oak tree transporting you to the coronation of England's first king (925AD), every

[14] Give or take, depending on whether a storm has washed any away since writing this.

inch below the surface of a seagrass meadow is a real-time window into history. The years simply spill behind you the further you look – our starry sky below the seas.

Before you slip, consider the fact that this topography could one day save us from drowning. A mind-bender of a study published in *Nature* (2019) discovered that these layers of carbon below a seagrass bed in fact *elevate* the sea level. The thought is a raised seabed would soften the blow of climate-induced sea-level rise – a genuine threat that many coastlines will have to confront within the next century. How much would it cost to raise the seabed artificially? Don't even.

More than that, a research team in Spain in early 2021 found strong evidence suggesting that seagrass meadows may help to hoover up the plastic waste entering the sea. The seafloor is increasingly cited as a final resting place for terrestrial plastic waste. It's not a good look. Up to 900 million items of microplastic (pieces less than 5 millimetres in size) *every year* are thought to be collected and bundled into 'Neptune balls' – in the Mediterranean alone. Around the size of a golf ball, the researchers estimated each kilogram of these planty, fibrous Neptune balls can store an average of 1,500 pieces of plastic. Incredibly, it seems as if this natural trapping process halts the plastic in its tracks – *preventing* it from entering the open ocean. When washed ashore, these bizarre formations have a similar vibe to a hairy coconut.

Although it's likely the plastic offers zero benefits to the seagrass, any harm done to the meadow remains to be discovered. And I can't help but marvel at this habitat's almost altruistic peace offering to us – a gentle reminder that we ought to be more careful. Another dimension to its already insurmountable services. Seagrass: the great ignored philanthropist – what have we done to deserve you?

We were due to spend the night on the peninsula. Like my first-edition thoughts on departing Wales in 2003, I had mixed emotions as my flight with seagrass began its descent back to land. The top half of my body had physically checked out of the Irish Sea, and I couldn't remember what it felt like to have hands, but the bottom half wanted to carry on swimming over the meadow. A healthy balance, perhaps.

Lines of traditional cloddiau (ancient stone walls and earth banks marking field boundaries across the Llŷn) sketched our route west to the campsite, which lay a short hop from Morfa Nefyn. A wholesome affair, as several other tents and campervans dotted the lush meadow that overlooked the mirror of the sea. Dad and I drank wine from paper cups and had dinner in an empty pub garden down the lane. Towards sunset, leaving my shoes behind, I found a path that made its way to an outcrop of rocks on the edge of the headland. As I walked towards the sound of gentle waves lapping, the grass brushed like soft hair below my feet in a bizarre déjà vu. A shag settled on a molten sea. The sun lost strength below a mackerel sky, intensely orange in its last hurrah. I appeared to have stepped into a guided meditation. Exercising a similar intention, a family of four stood with their arms around each other, looking across the horizon. I offered to take their photograph, and we exchanged niceties about the weather, the silence and the times. And they didn't ask why I wasn't wearing shoes, which was a relief.

In the wake of Project Seagrass's success in Dale, I had asked Evie what she was afraid of. I enjoy asking people this. It takes a certain level of tenacity for someone to dedicate their working life to protecting something they care about. Something they know they might have to grieve for. Passion knows no age and, although young, Evie's bond with the sea runs deeper than for most. She spoke of her 'terror', when imagining diving in a sea and seeing bare sand. The irony of a paradise lost, when many of us would see such underwater minimalism as something straight out of a travel brochure.

With seagrass meadows supporting over a fifth of global commercial fish species – and home to 40 times more marine life than a bare seabed – the economics are as clear as the water itself. Cod, plaice, pollock – the ones on the chalkboard at £7.60 – over one-third of these British commercial fish rely on the haven of these meadows at some point during their life cycle. Mothers and fry seek safety in a palace of fronds, joining sponges, prehistoric worms, anemones, cuttlefish, squid, snails – the bedrock of the ocean exists between these blades. And, according to Evie, 'This is absolutely the number one benefit of restoring seagrass.' I'm inclined to agree.

But, capitalism prevails. Countless papers cite the immense food and financial security afforded by fisheries as the most persuasive argument for seagrass today. In Wales, a lengthy debate continues between conservationists, fishers and the Welsh government – rowing over increased access granted to artisan scallop fisheries around Cardigan Bay. And I do understand the challenges here. Just the other day in the dawn of 2021, as I write, a headline from *The Times* tells us how trawlers are damaging most of Britain's protected areas. But how odd that it seems to injure the very habitats that enable the catch. Not just the catch of fish, of course, but carbon, as we now know.

Nature Climate Change published a reel of research in 2016 that, to be honest, scares me. The act of damaging seagrass beds and other marine vegetated habitats, be it by storms or us, can *release* its stored carbon into the atmosphere – waving it into the sky like a helium balloon. Such disturbance is releasing one picogram of (once buried) carbon annually. In case you were wondering, that's

1 trillion kilograms. Or 1 trillion bags of sugar. Either way, I lament for logic.

Here's a beautiful thing, though. Seagrass affects no one, unlike many of the other large-scale restoration movements being proposed across Blighty (wetlands, woodland, peatland, unimproved grasslands). Perhaps this is what's spurring further seagrass restoration projects around the country – including England's largest, in Plymouth Sound. The only interaction with intensive agriculture it will have is mixing with its pollutants. Its rendezvous with humanity being either via a propeller blade, anchor or inquisitive wanderer. 'There is just no conflict!' Evie stated triumphantly. 'From the start, the areas of meadow restoration have remained in full use to fishermen. They understand that this meadow will only *increase* their catch, their livelihood. It's just a winner.' For everyone.

It was my last morning before home time. Dad had brought his paddle board with him – meaning a final flight across the meadow was unexpectedly on the cards. I was curious to explore deeper into the bay at Porthdinllaen and view the fields from above. Perhaps they would remain just out of sight, but there was no doubt they were firmly on my mind. Others – paddle boarders, kayakers, swimmers, walkers, readers – were finding something else here.

Skimming on top of water untouched by wind, re-entering this world from an aerial view felt an almost priestly endeavour. More visible than the previous day, the mountains of the north added a rugged quality to this softly lit bay, which I enjoyed very much indeed. Various sizes of moon jellyfish offered an escort back to the main part of the meadow, making me feel important. Their rings of pink gonads (reproductive organs – at the bottom of their stomachs) stared up at me like freshly rolled dice across a fluid poker table. I wondered whose roll it was next.

Once again, the seagrass bathed in its aura below an overcast sky. I shifted from standing to lying on my stomach, one arm dangling in the water, face glued to the moving window below. Every so often, a lip of water would envelop my nose and allow me to exist in both worlds, just for a moment. Swaying with the motion of invisible currents, it was like the meadow was workshopping a new samba routine and wanted my (completely uninformed) opinion.

Circling an outcrop of rocks that hosted a mothers' meeting of herring gulls, I laughed at their noisy fluster when they were occasionally mobbed by a great black-backed gull that jostled for the best perch. A grey seal, the harbour master – was more present that morning. His dog-like head surfaced at regular intervals, outlandish whiskers X-raying my intentions and disappearing back to record them in his logbook. It is times like these that Evie hopes will feature in the national vocabulary. 'As soon as you see seagrass for yourself, you wake up. I've seen so many people then say, "Oh, my God, we're actually losing this", and then they want to help.'

I can't help but wonder how tempting it must have been for those involved in Project Seagrass to sow in secrecy. Admittedly, I once hammered a bird box to a tree in the local park just to see how unforgivable being such a guerrilla might be for the local council. Three coal tits fledged last spring. There is an undeniable appeal in abandoning the frustration of formal environmental decisions and just cracking on with the job yourself. Sure, armed with the knowledge and workforce, the Project Seagrass teams could have easily dumped a bunch of seeds all over the British seabed, and we would be none the wiser. Yet if the melody of the sea is to be truly resolved, human community must be at the heart of the song. Better to act in full frontal view of the public and give them a chance to be curious – given that human understanding is the key to survival for all habitats. And this celebration of the very best of humanity is what has made Seagrass Ocean Rescue such a roaring success.

Thinking about it, the power play of a plant is something I'm unsure we will ever truly understand. Will we ever get over seeing plants as a boring affliction? A handicap to modern life? Maybe. Much of ecology supports the '*Field of Dreams* hypothesis', following a 1989 American baseball film of the same name that had Kevin Costner hearing a mysterious voice telling him that if he built it, they would come. Far from marking a baseball diamond on the seabed, I do like the notion. Restoring habitat is the first move we should make – our *Queen's Gambit,* if you will. Using plants to rebuild biodiversity and end this game we find ourselves playing on the cluttered chessboard of the world. 'If we build it – they will come' should be the mantra engraved onto our conscience.

Time to flick the clock in favour of the ugly duckling, don't you think? But first, home. Laundry, a hot shower and finding a bat being the main items on the list. Checkmate.

CHAPTER FOUR

Grey Long-eared Bat

I read somewhere that a common misconception about a tree's root system is that it mirrors the trunk and branches above. In reality, roots are surprisingly shallow, dominating only the upper 60 centimetres of soil, instead opting for a widespread approach – fingers in lots of pies. I understand. Something about Devon's sandy, pebbled and clay soils has allowed my own roots to grow far and wide. I'd like to keep it that way.

For trees, their health depends on various conditions, various *elements*. And for me, regular raids along the South West Coast Path, England's longest continuous footpath (spanning 630 miles), offer a unique elemental hit – tapping my roots in a little deeper. Two 300-mile hikes along the Devon and Cornwall coastlines bookended my early 20s. Four-season sunrises and sunsets have taken me hostage many times. Weekend snoozing and gossiping among thrift on the headland with my mum – 'our headland', we call it – permeates the summer albums. All manner of trivial frustrations, laughter without cause and raw contentment have been released onto this trail over the years, and I've found it a better listener than most. My dad tells me, 'The best views are often behind you.' But standing on the track I know so well, with a blue-and-copper-coloured Atlantic waving me forwards and a bee investigating my purple top – I begged to differ.

It was late July, and I had recently returned from Wales. Screens had swiftly replaced seagrass meadows, and I knew my next adventure would play out on home turf. To be honest, I revelled in its less chaotic itinerary. My original plans included being ever so adventurous and sea-kayaking the route, paddling parallel to the coast path. However, large swells and unpredictable currents led a coastguard friend of mine to advise against such absurdities. So, I did what I knew best and pulled on my boots. To share the experience, my friends Harriet and Briony joined me for the day. We met while working in the local outdoors shop together a few years ago and quickly decided to socialise beyond the stock room with immediate effect. The truth is, I've found it hard to discover women my age who consider a decent hike along a shapely section of Jurassic Coast a 'good time'. These girls are a rare (but not endangered) breed.

On the whole, there remains a significant difference between going for 'a walk' and going for 'a *walk*', if you know what I mean. But here were three such humans, booted and backpacked, ready to go. Whether my friends were keen to stick around into the night for a bat that almost guarantees a no-show was unlikely, but it was summer and the day was ours.

We were heading for Seaton, 20 kilometres east of our starting point above Otterton. Conger Pool, Sandy Cove, Little and Big Picket Rock, Tortoiseshell Rocks and Chit Rocks are stationed like stepping stones en route. Awkwardly nestled between charming Beer and historic Lyme Regis, Seaton is a seaside town that, bless its heart, has tried and failed. The plain middle child. My old secondary school is nearby, and despite the nearby (*excellent) Seaton Wetlands, unusual flora and wading birds were not on the agenda in my late teens. Instead, the day a Tesco Extra arrived complete with a Costa was the end to any productivity in our free periods. Seaton became our forgivable concrete escape – and after all this time, I was looking forward to the rendezvous.

Although Harriet and Briony were competent walkers themselves, I decided not to fully reveal the 'challenging to severe to strenuous' rating of the walk to them. The days leading up to the hike had been sweltering. Our hike hadn't even begun, and already the humidity had risen to a crescendo before settling over our heads in a pregnant cloud. With the total absence of wind, we were hot, sticky, but excited to be out together. Resuming outings with friends was a bizarre novelty following four months of social restrictions across the UK.[15] Ladram Bay lay ahead of us; its iconic red sandstone stacks a gateway to the caravan park that sprawls inland. A few posh houses sensibly hid behind a screen of Scots pine. Hundreds of boot prints dimpled the track, skirting around molehills and cowpats. Many others had been here before us, this section of the path between Otterton and Sidmouth being as popular as tea in a crisis.

I often forget that the Jurassic Coast is a UNESCO[16] World Heritage Site, meaning its stablemates include: the Great Barrier Reef, the Taj Mahal, the Grand Canyon. Oh, and the Tower of London. Hey, who knew that the UK has 32 of these?! Thirty-two areas oozing with 'cultural

[15] Up yours, Corona.

[16] United Nations Educational, Scientific and Cultural Organization

and natural heritage … of outstanding interest … to be preserved as part of the world heritage of mankind'. I'll certainly bear that in mind, then, next time I'm in someone's armpit on the Circle Line en route to said Tower of London. But stand with hands on hips atop the Jurassic Coast's High Peak, and it takes but a few seconds to justify such an accolade. At 157 metres above sea level, on a clear day, it's possible to see the Jurassic Coast in its near-entirety. Had we been walking towards Dover around 450,000 years ago, I may have chosen a 20-mile wander across an icy chalk ridge to Calais for a croissant instead.

The unearthing of flint hand axes and Neolithic pottery shards on the Jurassic Coast revealed two periods where High Peak was significant to our ancestors: in the Stone Age (around 4,000BC to 2,000BC), and then again in the interim between the Roman retreat from Britain and the West Saxon occupation of Devon (roughly between 400AD and 700AD). I wonder, will our generation's archives preserved in 'the cloud' be as interesting to dissect? Or will it just be: '*What I Ate*: a series'?

The Jurassic coastline officially begins in Exmouth (tried, failed, tried harder, and now rapidly improving seaside town) and celebrates a 95-mile geological banquet of coastline, before ending at Old Harry Rocks, near Swanage, in Dorset. This coastline remains England's only natural World Heritage Site. Safe to say, walking across its cliffs feels (at times) like you are gatecrashing the most outrageous film set. I half expected someone to shout 'CUUUT!' at any moment.

Three huge chapters of Earth's biography are crammed into these 95 miles. A stuffed crust, so to speak, that is easily more than 250 million years old, yet far from stale; stretched and baked into deserts during the Triassic Period, around 252 million years ago. This blazing, sandy time marked the beginning of Earth's 'Middle Age', as well as the dawn of the giants that pervade our wildest glances into the past. For this Mesozoic Era was also the 'age of dinosaurs'. A few million years later, sea levels rose, and the desert upgraded into a tropical sea during the Jurassic period. Closer still and the Cretaceous period cycled through sea levels, buried and stored forests, tilted rock layers in favour of the east and eroded the west before finally assembling the sandstone and chalk we see today.

The rockfalls and landslides that often headline the news around here continue to reveal the life that lived on these ancient lands. Mary Anning's humble curiosities transformed fossil-hunting from humanity's diversion into a great global excursion. Born in 1799 in Lyme Regis, young Mary took to beachcombing, hoping to sell trinkets to boost the family's poor living conditions. At the time, the Napoleonic Wars were heating up, Jane Austen had just finished *Sense and Sensibility,* and the 'theory of extinction' had only just been introduced. Assisting her father in fossil-collecting was a most unseemly pastime for a Georgian girl, but Mary had a bit of a knack for it. A 5.2-metre-long skeleton of a marine reptile (an ichthyosaur, 'fish lizard') and a complete skeleton of a plesiosaur (another dizzyingly large marine reptile) are among her most astonishing discoveries. Dying of breast cancer aged just 47, Anning's was a short life devoted to the pursuit of knowledge. Her legacy is still revered to this day and she advanced our understanding of the Mesozoic, probably more than any other individual.

Back on the coast path, we were making quick progress. Herring gulls rode the breeze, cruising at our level, the syntax of their muscular bodies suddenly making sense away from the bins and flustered picnickers on the nearby promenades. We ascended Peak Hill, which levelled at the borders of Mutter's Moor, an area of busy heathland offering one of the best views in England: a quilt of farmland sprinkled with sheep, stitched by villages, church spires resting among enormous nests of bracken, woodland and pebblebed heaths. With good visibility, you can easily see the crest of Haytor rising on the swells of Dartmoor. Gorse bushes were strung with their flowers like Christmas lights. The sea, as ever, at the cusp of it all.

Every time I come here, I recall a moment during a family walk many years ago when we shared this view with a man and his dog. Only when the man asked if we would be kind enough to describe it to him did we realise he was blind. Six-year-old me was confused at how someone who could not see could still so clearly appreciate the beauty of a landscape. But time has taught me that, when outdoors, we simply have to close our eyes to remember that sight is only a fifth of a feeling. Given a chance, our other senses are more than willing to chip in and fill the edges. I'm still sure that the man saw more than we ever did in that moment.

During 2019, I was involved with a conservation project called Back from the Brink – a collaborative initiative involving several organisations. In a spectacular pool of resources and people power, more than 4,000 volunteers across 19 concurrent projects shift vital gears. Site surveys, monitoring, public, landowner and local council engagement – Back from the Brink champions an agile approach and demonstrates the connectivity it hopes to restore. Target species range from the willow tit (the UK's most threatened resident bird), narrow-headed ant (exactly as it sounds), to the grey long-eared bat (keep reading). And it was delving into the private life of this grey long-eared bat (*Plecotus austriacus* simply rolls off the tongue, does it not?) – the bat that disobeys evolution as it whispers across vanishing meadows – that made me want to hike 20 kilometres on a boiling July day in the hope of seeing it again. I'm bothered that 97 per cent of the unimproved grasslands (AKA wildflower meadows) over which they like to hunt have disappeared since the Second World War. But, luckily for me, Devon remains a bit of a pivot-point for several species of bat, as remnants of this habitat linger in the shadows.

I had arranged to meet Craig Dunton again, lead bat project officer for Back from the Brink and Bat Conservation Trust (BCT). On the whole, summer is a crucial time for bats, as pups are born and command the feeding frenzy. It's the best time to try to see them, actually. They are more visible around the countryside between April and October while they rear their young before they retreat to hibernate overwinter in caves, cellars, disused mines, and such. Craig had planned to revisit and survey a farm near Axminster, home to one of England's remaining eight known maternity roosts[17] of grey long-eared bat. I fancied myself a bit of this, and on my calendar, I had marked that evening: *Meet Batman. Become Robin. Find bats.*

Before we go further, I just want to be clear: we humans have an appalling relationship with bats. The bat's reputation has not aged well. Yes, we rightly connect bats with the dark, but because society has taught us to fear the dreaded night, any regard often ends there – and it has been this way for some time.

From around 100BC until the present day, the vocabulary we choose to refer to these animals seems as limited as our understanding: 'disease', 'vampires', 'blood-suckers', 'demon', 'death', 'hell'. Ancient clay statues haven't exactly

[17]Where the bats live in colonies, give birth and raise their young during the summer.

helped. Some dating back to 300AD in Central America depict 'Death Bat', a human-sized vampire bat, the demon of the underworld, as he resides in his blood-soaked cave with his demonic army. Dear Aesop and his fable *The Birds, the Beast and the Bat* taught readers to reject and misplace bats as nothing more than an anomaly – 'He that is neither one thing nor the other … being both bird and beast.' Slavic folklore provoked believers to associate bats as dooming corpses to a satanic vampire afterlife. And, lest we forget, there was Bram Stoker's *Dracula* – the blood-thirsty aristocratic projection of Victorian anxieties. Closer to home, millennial vampire fiction tempts viewers to desire, sexualise and fear immortality. We're asked to crave and detest the supernatural and animalistic, all in the same breath. The only paradox I see is our famine of sight. Blind as a bat? Hun, please. More like blind as a human.

But not every culture holds bats in such low esteem. Where Western nations primarily see fear and superstition, China sees happiness and good fortune. Chinese artists have depicted the 'five blessings' of health, long life, prosperity, love of virtue and natural death – each as a bat. Bats, of course, are well known to science and society as vectors of deadly viruses – but again, hold that thought if you can. The negative bias many of us harbour is essentially a choice with which we have become comfortable. We borrow the traits we need for fancy-dress, and apart from an occasional pop-culture article celebrating a bat's fascinating biology, it's a real pity.

As a group of animals, bats have outlived people around 250 times over, Earth being their rightful pad for more than 50 million years. Around 1,400 bat species have emerged during that time, making bats the second largest group of mammals after rodents, comprising about 20 per cent of all mammals. It's easy to overlook that bats remain the only mammal that has evolved full flight, already achieving what we never will. Sure, flying squirrels take to the air, but only for short glides between trees. And sharing more DNA with horses and dogs than with mice, bats render the 'flying mouse' idiom delightfully obsolete. We, too, have a place in that vast family tree alongside bats. Teeth, fur, and breasts that lactate in response to the hunger of live young comprise a mere fraction of our mutual code.

Bats form the group known as Chiroptera – a Greek word translating to 'hand-wing'. My friends, mammals taking to the air is part of the same glorious conundrum that put whales into the sea, gave dinosaurs feathers and drew them to our bird feeders. In the spirit of Mary Anning and her beachcombing, 50-million-year-old wisps of bone and teeth have tumbled into the bat fossil record around the world. Still shrouded in quintessential mystery, it appears bats evolved to glide before full, flapping flight – every millennium allowing an extra window of membrane to lift them higher into the sky.

It's only natural that a group of animals so numerous would come in all shapes and sizes. One only has to take the midnight train from Bristol Temple Meads station to draw parallels with our species. The wingspan of bats ranges from 140 millimetres to more than 1.5 metres. Most eat insects; some eat fruit or nectar. Others prefer a little bit of everything. Wikipedia tells us the grey long-eared bat veers on the large side. Granted, its huge ears and taut, furry abdomen give it a full-bodied, robust appeal. Like a fine wine, perhaps?

We have another species of long-eared bat in British skies – the brown long-eared variety. Although both are known, rather wonderfully, as the 'whispering bats' (we'll get into this), a couple of crucial distinctions are important to note. Craig (our certified Batman) spoke of the stark differences between the numbers of each species – there being around 1,000 grey long-eared bats to 900,000 brown long-eared bats in Britain. He also mentioned how much their range varies, with browns found right up into Scotland and generally being the more adaptable of the two. Put simply, the grey long-eared bat is pretty much as rare as you can get. Being a southern European species with the most decent populations on Spain and Portugal's Iberian Peninsula, the tiny numbers that remain in the UK are particularly tentative, as our shores define the very northern edge of its range.

With this in mind, grey long-eared bats have all the usual paperwork: officially 'protected' in the UK under the 1981 Wildlife and Countryside Act and a European protected species. Long-lived for a mammal of their size, it's not unusual for some to live as long as 10 years, and the oldest on record was 14 and a half years old, and thus they remain one of the most valuable thermometers gauging the health of our landscapes. The devout onslaught of intensive farming, token pesticides, associated insect decline and modern development has confined remaining UK colonies almost exclusively to the south and south-west of England. As is becoming the yarn, our current knowledge is barely doing the actual lives of these species any justice. With the grey long-eared bat especially, much remains unknown.

Don't get me wrong, but I think we can agree that (like all bats), our grey long-eared was not blessed by the beauty stick – at least not by modern standards. Small beady eyes, pug-like face, frankly *massive* ears and clawed, membranous wings certainly aren't represented by the same publicist as the dormouse. You know, I think somewhere along the line, centuries ago, we confused the word 'weird' with 'ghastly', 'eerie', 'unnatural', 'grotesque' and 'freakish' – all valid synonyms in the Oxford English Thesaurus. Such words led us firmly down the dark alley of digression, far away from the character of this unique animal. I think it's high time that we return.

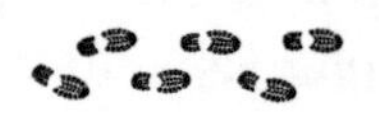

I had missed my girlfriends. The endless chatter. How often *should* we wash our hair? And do you face towards the showerhead or away from it? Is going to bed at 9.30 p.m. considered sad or sensible? We spent a good portion of the stretch between Sidmouth and Weston Mouth riding on circular sentences of 'should they?', or 'shouldn't they?', of 'does he like, *like* me in that way though?' and debating the ethics of the girl making the infamous 'first move'. A bench that was dedicated 'To Brian. Who loved this view' – was gradually being reclaimed by grasses. Grasshoppers chirred in the breeze. I felt like we were on holiday, falling into the trap of associating the exotic and interesting with foreign shores. Lush valleys and old, weathered woodland folded into a cleavage between clifftops. Everything had a hardy, steadfast exterior.

Our quads endured a battering on the ascent of Weston Cliff. From all directions, its steep climb is infamous. A summit so long and flat, it looks as though it's been lopped with a cheese wire. The South West Coast Path has an uncanny ability to test your emotions. One minute you stand empowered and elated, having just reached a peak; the next, you slither down the ladder to the shore. This game repeats itself along the entire 630 miles. And it was at Weston Mouth beach that Harriet, Briony and I stripped down to our underwear and ran (*drunken-looking stumble), squealing, into a brown sea. Despite the cold rush making us sound like a full maternity ward in sure need of hot towels and a midwife – it was bliss. Rinsed by the waves, the breeze stepped in for the blow-dry. Salty, refreshed and proud, we dressed, scoffed sandwiches and rejoined our path to the grey long-eared bat.

It's to be expected that humans would look to the natural world for a little inspiration. After all, most plants and animals have spent a bit more time on Earth – innovating and finding solutions. From Chinese inventor Lu Ban observing how lotus leaves deflect heavy rain to Da Vinci's flying machines and a kingfisher's dive teaching Japanese bullet trains to fillet the wind, the umbrella, the plane and the train have enabled humans to progress at astonishing speed. 'Echolocation' – the ability for animals such as bats and cetaceans (hiya, harbour porpoise!) to produce, receive and process the echoes of ultrasonic pulses – has been a trait that we have eagerly added to our little syndicate. From nautical charts mapping the seafloor to instructing warfare or betraying a hair grip on an airport security scanner, sonar makes a vital contribution to our world.

It's thought that echolocation evolved separately in bats and whales, in what science refers to as 'convergent evolution' – where evolution generates similar solutions to problems in totally unrelated animals. The problem was simple for both the whale and the bat: finding food in the dark, whether on land or underwater, is a total bloody nightmare. And a landmark piece of research in 2019 found identical mutations in the set of 18 genes associated with mammalian hearing – in both whales and bats.

This beautiful coincidence unleashed the enviable skill of these utterly distant relatives to employ sound waves to navigate, communicate and hunt. It seems that the earliest bats, Old World fruit bats, flew first and developed tongue-clicking detection a little later. Scientists now believe this to be a rudimentary manner of using sonar. Further studies point to echolocation evolving in bats multiple times in response to environmental challenges. Our 'whispering' grey long-eared bat has acquired a suite of elite skills to outwit its prey as it gleans a busy meadow overnight. Detecting its low short harmonies following an exhausting trek across one-third of east Devon was going to be a challenge.

Such immaculate rearrangement of genetic code in response to ecological need is more likely to arise in groups of animals that have a healthy 'gene pool'. Now, on the whole, I cannot wrap my head around genetics. It presents a similar challenge as looking at algebra without crying. But, all we need to know here is a fundamental distinction: a 'healthy' gene pool simply means an assortment of different genes floating around an interbreeding population. Where having children with someone who isn't already in your family is usually the preferred choice.

In contrast, an 'unhealthy' gene pool is the opposite. Stunted growth, poor cognitive function and reduced fertility are some of the red flags that fly in a population where your brother might be your lover and father of another mother. (European royal families have somewhat perfected this over the years.)

The point is that genetics shape our armour against change – from climatic shifts to habitat availability. The more genetic variability a species has on offer, the more likely useful genes will be chosen and inherited by future generations. More genetic variation simply provides more space to play with – more room for advantageous combinations of mutations to arise in the genetic make-up. It's just a better world for all. This is Darwin's 'natural selection' in a nutshell. For many endangered species, however, it's not so easy any more. These vital pools are shrinking – and it's looking like the gene bank of the British grey long-eared bat may be nearing its overdraft.

Dr Orly Razgour is one of the world's leading experts on bats. A colourful academic portfolio has taken her to Africa, the Peruvian Amazon, Europe and the Middle East. Our grey long-eared bat became the focus of her PhD at the University of Bristol, aiming to address the gaping hole in our understanding of this elusive mammal. Now a senior lecturer in ecology at the University of Exeter, the research

Orly leads with her team is of international significance. Orly is also effortlessly cool. She likes coastal walking, camping, and dogs. Chatting over Zoom, I quickly clocked her funky earrings and striped backdrop as things I wished to adopt. It was a rare occasion in that she was exactly how I imagined her to be – if a little more brilliant.

Orly spoke of how the core range of the grey long-eared bat population roosts around the Mediterranean Basin and the Iberian Peninsula. 'This is where the species has the highest levels of genetic diversity because these areas remained climatically suitable for them for hundreds of thousands of years,' she said. Here, the bats had safe harbour in a forested life afforded by Mediterranean climes, while waves of glaciations gripped northern Europe during the last Ice Age (around the time we could have walked to Calais for that croissant…). 'They thrived here for many generations,' Orly explained, 'and as a result, developed a very complex population structure and high levels of genetic diversity.'

As these southern European populations of grey long-eared bats burgeoned their gene pool like this, their British cousins could only recolonise when the ice retreated in the last 5,000 years or so. Because of this shorter evolutionary history, Orly described how modern British colonies represent a small fraction of the genetic variation found in the Iberian source population. Lump that on top of recent, severe population declines – and we've got ourselves a bit of a problem. 'Low genetic variation means lower adaptive potential,' Orly said simply. In essence, British grey long-eared bats have had less time to fashion their genetic chain mail, leaving them vulnerable to attack.

Orly led a study in 2017 at the University of Southampton, teaming up with Professor Gareth Jones, who coincidentally was one of my lecturers during my undergraduate days. Trialling a novel method, Orly and Gareth studied genetics and ecology in tandem to observe whether the genetic make-up of the grey long-eared bat can influence its response to climate change. Comparing the Iberian versus southern English populations lit a fascinating candle of thought, which applies to all species: are they physically equipped to deal with the planetary changes that are to come? Do all, or do any, species possess the genetic stamina to keep up with us, the merciless pace-setters?

Well, in the comfort of the short term, they might. Orly and Gareth used their data to predict that by 2080 the British Isles (good old Blighty of all places!) might become their safe, climatic haven, following an inevitable soaring of Iberian temperatures. As a rule, ecologists widely recognise Mediterranean ecosystems as facing a monumental biodiversity shake-up in warming years. Already, the average Spanish summer lasts for five weeks longer than summers lasted in the early 1980s. And joint research by Spain's national weather agency and Ministry for Ecological Transition linked this change to around 1,200 annual human deaths. During the same period, the Mediterranean Sea has warmed by 0.34°C every decade. With the exception of north-western Atlantic coasts, Orly writes how 'all Iberian populations are projected to experience maximum temperatures outside the current thermal range' of the grey long-eared bat. I suddenly feel quite hot.

Orly's research predicts a turning of the tables within decades. A north-western shift in the distribution of suitable conditions will uproot the grey long-eared bat from the Mediterranean Basin, pushing it towards our shores. So primed was I for a bleak outlook, I found this surprising. Orly read my thoughts. 'On the one hand,' she said, 'this *is* positive because we may even see this bat come up into Scotland, but on the other …' – I could tell from her sigh she had delivered this line too many times – 'what we've got to remember is that with the predicted loss of the Iberian populations, we will be losing this species' highest reserves of genetic diversity. That means we may lose their evolutionary potential and their potential to respond well to environmental change.' Sustaining a population whose gene pool has evaporated below the crucial level is as frantic and futile as a kettle boiling without enough water. If echolocation emerged from ancient reservoirs of undisturbed genes, I dread to think of the spells unable to be conjured in our bats of the future, the secrets and solutions that we will lose.

Despite the significance of Orly's research, it remains unclear whether the bats in the Mediterranean will be physically able to relocate when rising temperatures overwhelm them or whether they're going to be stranded and could disappear. Of course, we always need a better understanding of the ecology of these animals and predictions of their survivability under an unstable climate. But I wonder how lenient the clock is, since we have already had ample opportunities to learn more. How many penalties are we going to deliberately miss?

Yes, the thought of more bats with huge ears mapping British skies is a tantalising prospect, yet Orly was quick to reiterate the long-term lack of genetic resilience in British populations. She suggests that a mammal with the biological sensitivity of the grey long-eared bat may not withstand climatically altered areas in the future. So, with the British Isles already at the mercy of erratic weather, and if we're estimating these animals' physical and genetic endurance, long-eared bats currently look more suited to a sprint than a marathon. I picture a sign hanging from a dangling carrot promising *Shiny Renewed Prospects!* in balmier Britain for the grey long-eared bat – only for those prospects to be dashed in a few years because somewhere down the line, the climate caused their biology to fail.

Orly concluded the study by proposing the need for an 'evolutionary rescue in most vertebrates'. My mind raced. The very thought of this 'evolution' morphing into a (probably handsome and rugged) knight in shining armour, ready to slay all that dares to disrupt its battle plan, is reassuring. Not very feminist of me, I'll admit. Yet its fragility in the face of powers greater than it, greater than Earth's own religion, is frightening. Orly pulled me back. 'More accurately, this "rescue" is through gene-flow from populations adapted to warm/dry conditions into maladapted populations (ones adapted to cooler/wetter conditions) that will enable maladapted populations to survive as the climate warms up.'

Leaving Branscombe up its famed 'Heart Attack Hill', we repeatedly paused to catch our breath. When we turned around, our prize was a stunning view down the west of the Jurassic Coast. Walkers have two options to reach Beer: contour through a low, woody path around the cliff or wander along the clifftop. We chose the latter, not willing to let go of the height we had just worked so hard for. This section of the path featured old coastguard lookout towers, quiet meadows and allotments. The sun felt hotter than ever, coaxing us into a daze as we finally descended to Seaton's shingle beach. My friends had more pressing plans for the evening than waiting for an invisible bat. Understood. They disappeared in search of a bus.

With time to kill before I headed inland to Axminster, I lay on the beach overlooking a serene sea. The smell of fish and chips and gentle summer murmurs sent me into a sunburnt, weathered microsleep. With a long night ahead, I took my time to walk the last couple of miles inland, soon finding the track that led to the farm. A filmmaker and photographer called Neil was already there assembling a chaotic-looking camera rig to the side of an enormous old barn. Neil grinned wildly as the rig found its balance. Batman Craig soon followed. You see, the grey long-eared bat is so rare here that photos of it barely exist. Those that do exist leave much to be desired, given the challenges of nighttime photography. We've learnt most about this bat from studies in other countries and Orly's research. Neil's longstanding mission has been to capture some of the very first good photos of these animals in the wild – and at night. The farm itself is as idyllic as you might expect: a swathe of organic farmland in the bosom of the Devon hills, choosing a more mindful use of the land. Even from the car park, you can see – feel, even – meadows, woodland and ponds that sing. A choice to harmonise with nature here has transposed this song into a better key.

The huge barn was our base for the evening, its beams providing sanctuary for many bat roosts: noctules, pipistrelles and, of course, the grey long-eared. On the whole, bats are crevice dwellers. Mines, caves and underground environments are classic roosting sites, but to some extent, early farm buildings and cathedrals may have also accommodated bat populations. Wander through an old mill, barn or abbey, scout for little piles of moth wings on the floor, and you may be looking at the leftovers of a long-eared's dinner. With Devon being increasingly sought after by property developers, I had asked Orly whether modern housing threatens the already scarce roosts of the grey long-eared bat. I can't imagine that cavity-wall insulation, kitchen extensions and reroofing is as thrilling a prospect for this bat as it is for the people paying for them. Orly stressed the difficulty in forming any solid conclusion, but she worries about the lack of control these situations present. It turns out that science doesn't have quite the same authority as a JCB and temporary traffic lights. Funny that.

Craig handed me a bat detector. I had never used one of these before but felt reassured to have any assistance that might help me glimpse our whispering hunter. I was unsure whether any skill is associated with looking for bats, save from patience. But I approached the situation with similar vigour to when I asked Miss Bradford to promote me from Villager #5 to Angel #1 in the school nativity.

Our grey long-eareds call in low, short pulses anywhere between 30–50 kilohertz. Given that human hearing is most acute at 2–4 kilohertz, we need all the help we can get. The detector itself looks a bit like a Game Boy. Tuneable, a dial on the front sets the frequency of the bat call you're looking for. Then, a series of audible clicks, whistles and scrapes cutting through the white noise confirms whether your bat is present and echolocating. Dr Holger Goerlitz, a sensory biologist, was interviewed in a *National Geographic* article in 2010. I love his analogy when explaining the limits of a whispering call in a barbastelle bat. Like the grey long-eared, barbastelles enjoy eating moths with ears (hilarious image). Goerlitz spoke of how most bats' sonar offers sight like 'a bright torch', whereas a barbastelle's sonar, or indeed grey long-eared bat's, flickers more like 'a candle' – only lighting the immediate area around them in a radius of fewer than 5 metres.

Grey long-eareds will fly up to 5 kilometres away from the colony on a foraging trip, so Craig, Neil, a few keen volunteers and I were stationed around the main entrances to the barn to catch their entry and exit to the meadow. The light was fading along with the heat of the day. A green woodpecker chuckled in a nearby tree. The flowers of meadowsweet, lady's bedstraw, clover and buttercups perfumed the air. The main barn doors were open wide to the night.

Although the genetic fragility of the grey long-eared bat is (for the most part) beyond our control, we do have leverage over the health of its habitat. As vital to them as water, this insect-loving bat needs scrubby field margins and wildflower meadows in which to hunt. Not only were nearly all of these habitats sacrificed in post-war productivity, but associated insect biodiversity has also plummeted. More than that, a global review of insect numbers in 2019 revealed insect extinction to be eight times greater than that of reptiles, birds and mammals. If we do nothing, insects will become history within 100 years.

Without insects, many animals such as the grey long-eared bat will become starved of their food source. That includes us, indirectly, by the way. We must be careful not to forget the other animals that need the meadows as much as the bees. It may surprise you to know that bats are also highly effective pollinators. US researchers studying nectar-feeding bats write of the 'tequila connection', where Mexican long-nosed bats are solely responsible for pollinating all the agave plants that fix our Tequila Slammers – thanks to their furry bodies receiving a blanket of pollen as they visit flower after flower. In the UK alone, we drink nearly 2 million litres of tequila every year. The financial, let alone social implications of encouraging bat pollination are as clear as the spirit itself.

One of the main culprits driving insect decline is the amount of pesticides that we allow to shower the land. A recent enquiry found that 17,000 tonnes of chemicals have licence to saturate the UK countryside every year. As I write, there are talks of halving this by 2030 and pleas from environmental organisations to confirm the benefits of nature-friendly farming in legislation. Granted, this is all very well, but we forget how nature had already identified this problem for us – centuries ago. Supervising pain-in-the-arse species is one of the most ancient riddles that evolution sought to solve.

'Grey long-eared bats consume a lot of moths – and a lot of pests are moths,' Orly told me. 'Where they forage along "edge" habitat, it puts them close to agricultural areas, so the pests they eat are likely the ones directly affecting crops.' Grey long-eared bats hunt in both the forest and open meadow, and by default, they have an appetite for a greater variety of the pests that we wish to eradicate. Consuming up to 4,000 insects per night, the bat's appetite has encouraged some farmers in Europe to ditch pesticides altogether and instead plough the money they save into making the habitat more suitable for bats.

What puzzles me is that the pesticides we continue to administer can be responsible for the death of this natural pest management. A study looking at German bat populations and pesticides found that chemical residues can remain on flies, moths and spiders for up to two weeks. The bats that gleaned such invertebrates were contaminated, leading to long-term health problems. And if that's not sufficiently unsettling, emerging data show how pesticides from modern agriculture can linger indefinitely in the air – the silent, invisible assassin.

This battle for control should never have even happened. 'The impact of a loss of grey long-eared bats in the UK is not going to be immense,' admitted Orly, referring to their rarity. 'But in Europe, they are known to consume many agricultural pests, and this could be a huge service that the UK could miss out on because the population size is so small.'

Female grey long-eared bats have one pup a year – if they're lucky. I imagine that they would be pretty cute. Another long-eared bat study that Orly led (told you she was good!) found a fascinating connection between lactating females and unimproved grassland. Analysing data from radio tags on the bats revealed a preference for this particular habitat, especially while rearing young. This data immediately points to the value of wild grassland in yielding high-quality insect food to convert to milk for a hungry pup. It may surprise you to learn that up to 40 species of plant can grow per square metre of wildflower meadow, each fraternising with its insect counterpart. Safe to say, these meadows are among the richest habitats in the world, but now, one of Britain's rarest. Meadow connectivity is compromised, and that mother bat is going to have to travel further to find food.

You'll be glad to hear that Back from the Brink prioritises this as part of its epic tapestry project. Re-stitching broken habitats for these bats is motivated by the expected influx of settlers from a toasting Mediterranean. Managing the land for bats within 5 kilometres of the roosting sites is recommended to buffer their foraging territory. 'But it's not just a case of whether the habitat can be found,' Orly told me, 'it's whether the bats can actually *reach* it.' Ask a breastfeeding human mum about their energy levels, and they will tell you they are beyond exhausted. So, they're probably not feeling up to a messy commute to Sainsbury's. Producing rich milk that your offspring guzzles in minutes is all-consuming and burns up to 700 calories a day. 'If we don't piece the habitat back together, mothers are simply going to have to work harder, travel further and won't be able to return as often to feed their young,' Orly explained. I lament the unfairness of it all and secretly thank the universe that I wasn't born a bat in the twenty-first century.

Back to the night. At the barn, our bat detectors and weak diurnal bodies were up against a lethal combination of nocturnal intelligence, agility and wit. When grey long-eared bats take flight, research suggests that their huge ears can be angled to increase lift and help conserve energy. When they are roosting, their ears are curled back and tucked under their wings.

We are such strangers to the night. Every day, we rebel against its rhythm by switching on lights. We replace starlight with blue light. We disobey our internal clocks. Any time we now spend wrapped in night's cloak feels like reconnecting with an old friend we haven't spoken to for years. There is much to catch up on, celestial anarchy to settle. But I don't know anyone who doesn't love a starry sky. Fuelled by the same urgency as catching the sunrise, making an effort to be part of a clear night is still one of my favourite things to do. My parents sometimes woke my brother and me just to tell us the stars were amazing that night. It was always worth being disturbed. 'Look *deep*,' my dad tells me as we gaze up. And 'What were the stars like?' is one of the first questions I will ask someone who has just come back from a remote location (swiftly followed by requesting a detailed account of the menu for the week).

My fingers holding the bat detector were freezing. My bum ached from the hike to get there. There will always be something captivating about looking for a nocturnal animal. If anything, it presents a new rulebook in which to trust – waiting for dark to fall whittles away at our senses, our hearing and touch chisel to compensate for the loss of light. Our pupils dilated, like ink might blot a page. I stood at the back of the barn, straining my ears and eyes to map this other world. Barely visible, blending in with the wall next to the barn, Craig was in the zone. A sound like a squeaky wheel and smacking lips cracked through the receiver as something darted out of the barn. I checked my detector – 120 kilohertz – a lesser horseshoe bat. And so, it began. Craig mentioned that pipistrelles and noctules are usually the first to emerge and that if we saw a bat but didn't detect its call, it may well be a grey long-eared. There were eight of us in total. At regular intervals, we would correlate bat 'scores' huddled in the corner, as people at the pub might discuss last night's game. When a few noctules had flown over and some time passed, I reset my detector to the range of the grey long-eared.

Around 9.30 p.m., it was now prime time for our bat to begin its nightly forage across the meadows, gardens and field edges, sometimes leaving the roost for six hours. I like to imagine that if the barn owl considers itself nocturnal aristocracy, then the grey long-eared bat is bona fide royalty. Friends, this bat is elite in the very raw sense of the word. Highly manoeuvrable, its large, scaffolded wings permit both fast, straight flight across the open meadow, as well as accurate banking through cluttered woodland. Where a bird's breast keel forms a strong muscle base to control the wing beat, bats have one muscle, a flattened ribcage and a fusing of vertebrae. Not only does this just sound insane, but this minimalism also affords the bat agility and lightness that is unmatched.

As I mentioned earlier, a favourite snack of the grey long-eared is the yellow underwing moth, a regular resident of the meadow. A 'tympanic' moth, the yellow underwing has evolved its own set of skills to avoid being taken. The tympanic organ, a drum-like sensory hearing aid, has developed in response to the fear of being eaten by an ultrasonic hunter. Many moths have adapted in this way. Tympanic membranes can read a bat's sonar and sound the alarm bell – giving the moth a chance to evade capture. A group from the University of Bristol, including Dr Goerlitz (of the excellent 'candle' analogy I mentioned earlier), found that yellow underwing moths are tweaked so minutely to the sonar of bats that movement the size of an atom can activate the nerve cells in this hearing organ.

We've seen this before. From the gazelle slipping the grip of the lioness to bacteria outfoxing antibiotics, an invisible battle across nature strives for the upper hand. Nature's arms race. An impetus to become smarter, faster, stronger, more immune, yet as habitual and unnoticed as breathing. During our call, Orly described how the whispering bats could 'switch *off* their echolocation … to simply listen to prey-generated sounds'. The fatal betrayal of a wing beat and a rustle of antennae are sometimes all that is needed. The arms race between moth and bat has made the grey long-eared think beyond the standard-issue acoustic box, making it covert, sly and challenging to study. As primatologist Frans de Waal so rightly asked in the title of his bestselling book, *Are We Smart Enough to Know How Smart Animals Are?* Well, Frans, I throw my cap into the proverbial ring and hazard a guess that we are not.

If hair straighteners and various sprays, gels and mousses form a crucial part of your morning routine, you may consider yourself unlucky if your perfect mane gets caught in the rain later that day. Similarly, in nature, the more specialised you are, the more likely you are to lose out when a significant change occurs in your environment. Ecologists have seen Devon's bats foraging in suburban areas, but more on the margins of villages versus near the city. Craig mentioned how grey-long eareds are loyal to the maternity roost, often not even attempting settlement in more modern developments. They also roost-share with brown long-eareds fairly contentedly – a pleasant thought. Rows of ears huddled along a wizened oak beam like children on the side of a swimming pool, waiting for permission to dive.

Bats are a sure sign that an environment is healthy: their immense mobility can indicate areas of good water quality and resilient insect assemblages, and bats bolster our food security. We now know that grey long-eareds might stand a chance of surviving climate change in the UK if the habitat accommodates a healthy breeding population in the long term. Pesticides, hopefully, will take less of a toll on our countryside, but a mounting body of current research is investigating an altogether different kind of pollution. 'Electromagnetic radiation' includes radio waves, visible light, microwaves, ultraviolet and infrared. Bats are susceptible to such radiation, many of which govern their navigation, orientation and migration. And (albeit a little late to the party) it just so happens that our deliberate introduction of electromagnetic radiation – to help us navigate, orientate, and migrate within the nightscape – seems to sacrifice the very animal whose biology inspired it.

Jack Merrifield is an engineering and physical sciences PhD student at the University of Southampton. Aside from being a mate of a mate, Jack specialises in the effects of environmental pollutants on ecological communities. When I spoke to him over the phone, he was working on his PhD: mapping Southampton's artificial light and noise pollution in relation to bats and their prey. It soon became apparent that light pollution, coupled with the small headache of climate change, is an entirely different ball game. A new arms race may well be emerging.

'Light is wherever humans are, and it is having a hugely negative impact,' he told me. 'Climate change is a really important issue, but it cannot be looked at in isolation. We urbanise so rapidly … "money talks", right?' Even over the course of his PhD, the lighting spectrum of Southampton has utterly changed. LEDs[18] have swiftly replaced traditional halogen and halides in complex subtypes of 'cool blue' and 'warm white'. Southampton now has over seven different lighting spectra. Despite LEDs consuming around 50–70 per cent less fossil fuel energy than incandescent bulbs, lighting still (and will) account for at least 5 per cent of global carbon emissions.

[18] Light-emitting diodes (AKA bright AF)

In 2017, a team from the University of Exeter measured an increase in the range and intensity of human illumination by 2 per cent a year, leading to an erosion of seasonal light patterns and a glorious screwing up of Earth's biological calendar. From hormonal cycles to breeding to whether an animal gets dinner or not, the rhythm of night and day is the ringmaster (or at least it used to be).

'The impact on bats is just massive,' Jack admitted. 'The area around the lights create predictably clumped food sources – a kind of vacuum for midges and moths.' We've all seen the frenzy with which a moth fraternises with the lamp. And this mass exodus of insects attracted to the city glow that Jack described leaves those bats left in the surrounding countryside, caught in what some call an 'ecological trap' – where slim invertebrate pickings are all that's on offer.

'Crossing a noisy, illuminated street for many bats is like asking us to walk through a brick wall. It just doesn't work with its biology,' Jack remarked. Craig agreed, stressing how light-averse grey-long eared bats are. Artificial light can delay or even prevent emergence from roosts. Or cause bats to abandon roosts altogether. Research like Jack's supports calls for 'bat bridges' and 'safe bat paths', following similar logic to 'green bridges' over motorways, which safely corridor wildlife. Treelines, hedgerows, speed reductions, culverts, bat boxes and containment of 'light spill' are modest but promising thoughts to support the dwindling species of the nightscape. Bring on the day that a road sign hailing 'Bats are here' is designed and normalised across cities.

Connecting meadow habitats is vital, and a project by the Bat Conservation Trust (BCT) aims to ensure that. As I write, 'Return of the True Night Rider' aims to preserve the vital outpost of Devon's grey long-eared bat population. Not as I initially thought via a hot release at the cinema, but rather a joining of hands. Craig told me that landlords and communities are restoring crucial bat commuting routes between Devon and Dorset, 'An essential step in preventing the genetic isolation of colonies.' Jack agreed but insisted that more research and mitigation for artificial pollution is vital to their long-term survival. 'It's very rare that a bat will encounter just one negative impact at a time, so it's a question of looking at them *collectively*. Only then will you create a habitat that has low resistance to bat mobility.'

Through my receiver came a different call, like someone tapping a biro impatiently on the table or water dripping very fast – perhaps the Daubenton's bat. I had twiddled my dial, impatience tiptoeing around, tempting me to settle for any bat at all. But none of them was a grey long-eared. I dialled it lower, returning to 30–50 kilohertz: the white noise, our lullaby. If a grey long-eared bat appears and chooses to echolocate, then the noise we hear would be a faint purring. Apt for this particular bat, of course, to emulate feline superiority. Craig admitted it's rare to detect the returning bats. His admission suited my fading energy. I wasn't up for a six-hour stint. I needed my bed, YouTube and a bit of cheddar. So, we orientated our detectors towards the black hole of the barn, waiting for those yet to embark. Now nearing 11.00 p.m., it was utterly silent – save for the rustling of jackets and shifting of feet.

All of a sudden, two, maybe three, black shapes darted over me. I strained for any purring over my detector – any at all. Nothing. Unsure, I grinned up at Craig, trying to correspond through the thick of the dark, but I couldn't see him nor him me. The night sky got the grin instead. Our whispering hunter may have just left the building.

There was no doubt that I was having an exceptionally good night. Waiting for a grey long-eared bat to emerge and reveal itself is like waiting for a bus that may never turn up. But it's infinitely more interesting. As I write, the UK is in the grips of the most severe wave of the Covid-19 pandemic. Research around its origins is frantic. Reams of articles report how this pandemic has refocused our attention on bats as harbourers of deadly zoonotic[19] viruses, re-condemning them to their outdated cultural corner. This terrible, global press has triggered mass culling events in Africa, Latin America and Asia. An avalanche of journalistic prosecution. Photos of their skewered, barbecued bodies on sale in Asian wet-meat markets have thwarted their conservation – plunging them into a nightmare more haunting than the ones they inspire.

It strikes me that bats are a convenient reservoir for our problems. Even in our 200,000-year blip on Earth, our impact has been profound. And as a fungal network sprawls beneath the woodland, sometimes our hidden touches are the most influential. The way we talk, the language we choose can spell life or death for an animal as misplaced as the bat. This issue is so pertinent that Orly and others from the University of Exeter released several papers in 2020 dissecting the depiction of bats in light of Covid-19. And the conclusions are clear: these animals are terrifyingly mistaken. More time spent considering their unique biology and behaviour could be the ultimate teacher in helping humanity to prevent another public health crisis. 'Because what is really at risk here,' Orly implored, 'is that when we destroy bat habitat, bats come closer to humans and livestock. And that's when transmission can happen.'

[19]An infectious disease that is able to pass from non-human animals (usually other vertebrates) to humans.

'It's our impacts on the environment that are causing this risk. Rather than being an intrinsic problem with bats.' Because you see, it's all about how *close* we get to the bats themselves. How much we prod their habitats with our problems and test the boundaries. Knowing them is knowing our future. We could better predict how a disease might spread alongside land-use change. We could better predict for and mitigate its inevitable spill over into humans, livestock and other species. Knowing bats better could shift the baseline needed to future-proof ourselves and the natural world. But it's easier to slag things off instead of finding a solution, isn't it?

A study in 2017 at a medical school in Singapore found that while bats harbour a strangely high number of harmful zoonotic diseases compared with other mammals, the researchers stress that this does not provide grounds to fear them. Depressingly, in the same study, they report that 96 per cent of 'virological literature' snubbed the role of bats in the environment. Still, centuries on from 'Death Bat', we remain loyal to our epic oversight of these animals. What's new, eh?

'We need to focus on their ecosystem services,' Orly advised. 'We still need to be aware that they host many pathogens, but yeah – so does every other animal.' Our relationship with bats has outlasted that of dogs, horses and domestic animals. Jack reminded me how we once shared caves with bats: 'The secrets that we yield about bats provide unique insights to our own biology. And with climate change, we have the gift of foresight.' Good point. It's not every day with an issue as gigantic as this that we have the numbers, the data, the knowledge and tech to address it. 'If we don't take advantage of this rare position, I worry that this situation will just continue,' he said.

Look – it's complicated. It's off-beat. I get it.

I went to the let's-swap-bat-scores corner where Craig stood and asked the all-important question. 'Could be ...' he shrugged, looking skyward. He said while it's likely that they were grey long-eared bats, you can never be sure. That's good enough for me.

Towards the end of our call, PhD researcher, Jack told me how new evidence is emerging on some bats, finding how they can form friendships across species and actively travel further to maintain them – for years. I have so many questions – the research is unfolding as we speak – so I'll keep you posted. 'Even for humans to maintain a relationship for over a decade is remarkable,' he laughed. 'So, the fact that a small, aerial mammal can do it is fascinating.'

We were calling it a night. Packing away the detectors and helping to shut up the barn, Craig suddenly called to me in a loud whisper. 'It's poo! I think I've found – *it's poo.*' Jogging to a large table inside the barn, head torch on, he whipped out a bat poo ID guide as casually as though it were a set of car keys. On the table below the oak beams sat a small mound of brown, crumbly droppings. And they were *sparkling*. Glittery poo. Yaaasss! People, the grey long-eared bat defecates like a queen. A dead giveaway to its diet, the hard 'chitin'[20] exterior of the insects it eats are ingested, digested and expelled as a fine, fabulous powder within the faeces. It sparkled in the torchlight, like frost.

Photographer Neil and his epic camera contraptions didn't yield on that particular evening. But, I'm delighted to say that since that night (and after many attempts), he has managed to capture the first high-resolution night-time photographs of our English grey-long eared bat – and they are astounding. Ears seem larger than ever, eyes bright and seeking. I don't see a flying mouse when I look at them. I don't see vermin. I see a hunter – a highly evolved

[20] Basically, just a load of fibre that makes up for the insect not having a traditional skeleton.

piece of biological machinery. I see an animal that I want to know more about. I see an animal that deserves better.

Here, in this barn, the bats are free. Or, in Cockney rhyming slang, they are 'yet to be', which I prefer. A clumsy, exhausted smile broke my face. Glittering faeces at midnight is a lot to process. Plus, I needed rest. Because in two days, I was to embark on the longest journey of this rather wild pursuit – it was almost time for me to wend my way north. To Orkney.

CHAPTER FIVE

Black Guillemot

Just keep heading north, I repeated to myself. Being back on my bike was fun, but weaving through rush-hour traffic with a bulky and much heavier rucksack wasn't so much. I felt nervous. Exeter to Orkney is a mere lunge on the map, yet mid-pandemic, and it was little short of a mission via public transport. A 40°C 'Spanish Plume' heatwave smothered Britain that week. Moaning about sunburn, lethargy, and feeling 'hotter than the sun' seemed all we could muster in small talk. A grave error of judgement found me racing across London to catch my connecting train to Edinburgh, overdressed, on the hottest day in Britain for years. Costa del North Sea couldn't come soon enough.

With just 30 minutes to cycle from Paddington to King's Cross St Pancras, I felt tense. Diversions and road closures hindered my progress. I ended up weaving through back streets, hopping onto pavements and dodging roadworks near the London School of Hygiene & Tropical Medicine. All trains were operating on a reduced schedule, and only one train was Edinburgh-bound. I was going to have to pull a blinder to make it in time.

Beautiful people drifted along sunny pavements in floaty dresses. Negotiating my journey to King's Cross while regularly checking Google Maps, I watched* (*slightly envied) an exotic-looking lady I felt sure was heading to the Bulgari for a seaweed body polish by a beautician called Epiphany. I caught sight of my ridiculous reflection ploughing past a Prada window display, ribbons and couture ruffling in my wake.

Arriving at King's Cross with five minutes to spare, I hurried to get water for the journey, dropping my phone and tickets in a splash across the floor. In a scramble to shove my bike onto the first available carriage as the doors were closing, it turned out to be first-class. With the same urgency as if a Royal were about to arrive, I was ushered to 'move *quickly* along please!' I later realised I had been holding my phone with its torch on full beam the entire time, which explained why heads turned away as I blundered through the carriage – pah, who cared. Fuelled by the distant hope of a seabird on a remote speck of a northern isle, I was escaping.

A few years ago, my dad and I drove around Scotland's North Coast 500 route and hopped across a turbulent Pentland Firth to the Orkney mainland for a day. There was an aura to this archipelago that, even through dense mist and rain, captivated us. Ever since then, I have wanted to give Orkney more time.

Looking out the window as the train raced north, the land was flat and worked. Peterborough, York, Darlington, Newcastle. Lines of pylons and power plants faded into a mirage like an enormous chain-link fence, disciplining the trees and hedgerows that bordered farmland.

During previous trips, I formulated a list of things that may amuse the regular commuter. My favourite became playing Reason for Travel™, where you guess the purpose of your co-traveller's journey based on their clothes, snacks and rate of texting. Using these variables, my carriage alone yielded a mixture of: those fleeing a failed relationship, those escaping to a forbidden relationship, and those on a mission to rescue an investment bank from financial ruin.

Five hours from London, Berwick-upon-Tweed – a place I had only read about – brought a choppy blue sea to my right. Then clouds, moorland and crying gulls paved the way into Edinburgh Waverley station. Before long, I was eating a falafel kebab with a tinny gin in Princes Street Garden, ranking the spectrum of tattoos and fake tan in an adjacent group. I felt equally decorated with chain grease and blood stains around my ankles from the fight with my bike in my rush to catch the train, but also thrilled about my Edinburgh debut.

En route to Aberdeen to catch the overnight ferry to Kirkwall in Orkney, torrential rains and high winds dominated the following day. It seemed only fitting that in Scotland, I would arrive at the ferry terminal sodden. I could hear the roar of a rather, shall we say, *frustrated* North Sea. My earlier zest had faded, but I remembered the Robert Louis Stephenson[21] quote I had seen tactfully printed onto the revolving toilet door in the train: 'I travel not to go anywhere, but to go. I travel for travel's sake. The great affair is to move.' Agreed. Already my solo hopscotch across pandemic Britain on my way to distant isles *did* feel rather gung-ho – liberating, even – and I had many more steps to take before I reached my destination. So, I locked up my bike in the enormous hull of the ferry next to lorries

[21]The Scotsman who wrote the 1883 classic, *Treasure Island*. Originally named *The Sea Cook: A Story for Boys* – I'm glad to say that times have changed and it's a cracking read for all genders alike.

and freight, bid it a tentative safe passage and climbed on deck into the storm.

Lying about 10 miles off the Scottish mainland, at 59°N, around 70 islands lie in wait. Only 20 of them are inhabited by people, drawn to the stiff elemental brew of Neolithic history, culture, fertile soils, geology and wildlife these islands offer. It's a county close enough to the Scottish mainland to retain governmental influence but far enough to assume its own way of doing things. Ask anyone who has visited Orkney, and they'll either reel off an endless list of things they loved about it, or they'll be lost for words. It is unique. If you look at the whole of Orkney on the map (especially on one that shows the shipping lanes), it looks like a beech leaf in a later stage of decay. Brittle and fragmented, yet the veins still trace its form as it scatters pieces of itself into the North Sea. A eulogy to an oceanic jigsaw, and all within easy reach of the Great British doorstep.

Where the nearby archipelago of Shetland thrives on fishing, Orkney islanders farm. The blend of temperate coastal climes and varied geology curates rich soils – critical assets quickly recognised by Neolithic settlers back in 3700BC. I was shocked to discover that Orkney has the highest density of beef cattle per hectare in Europe, grazing 76,500 in 2017 alone. Anyway, back to 3700BC and equipped with seeds, livestock and the hope of building a new island nation, groups of farmers left mainland Britain and crossed the treacherous Pentland Firth to explore this curious new land. Together with the persistence of traditional hunter-gatherer know-how, the settlers were part of the wave of agriculture that defined the Neolithic period from 4000BC to 2500BC. It turned out that transforming their patch of land into something that provided reliably for the family offered a sense of ownership and sustenance that their previous nomadic lives lacked.

Such a renewed feeling of place gave settlers room to explore culture and community, giving rise to world-renowned burial sites, stone circles and monuments scattered across the islands. The Knap of Howar, for instance, dating back to around 3600BC, is a stone-built Neolithic house on the isle of Papa Westray. The architects must have done something right, for it remains the oldest standing building in northern Europe. In a similar vein and part of Orkney's UNESCO World Heritage Site, the unearthed village of Skara Brae on the west coast of the mainland is one of the most intact European relics of the Neolithic era. Following waves of rugged Norse raiders in the eighth and ninth centuries, Orkney was under the thumb of Norway and Denmark until vanquished by Scottish rule in 1472. The Norse ways still whisper across island culture, popping up in old sayings, seabird names and family traditions. Between all this, the various Viking expeditions to Orkney in the eleventh century – and the importance of Scapa Flow as a strategic naval base during both world wars – Orkney is dripping with history. You'd be a fool not to want to grab a straw and suck it right up.

Of course, it was wildlife that I wanted. Luckily, Orkney is genuinely one of the northern hemisphere's most popular nature hubs – a wild rhapsody for lovers and agnostics alike – and it does very well in the 'Area' department. Thirteen Special Protection Areas and six Special Areas of Conservation (SACs) as well as the odd Marine Protected Area (MPA) and National Scenic Area, all of which are well deserved given the place is heaving at its weathered seams with seabirds, seals, whales, dolphins, otters and the endemic (found nowhere else) Orkney vole. This is both a landscape and a soundscape, and the cries of wheeling birds take an equal role in a sonic trio with the wind and waves. Gales and low-lying glacial hills leave this archipelago almost entirely bereft of trees – it's really quite weird. A teacher I met there explained how schools often organise trips to the Scottish mainland, primarily to allow the local children to see and experience trees and woodland. I haven't decided whether I could live with that.

Birds gather in seabird cities across Orkney's sheer cliffs and above the productive seas that surround it. The most recent RSPB[22] census on the islands recorded nearly a million breeding birds at any one time. Such epic birdlife has prompted extensive research into the wildlife here, led mainly by the RSPB. Agonising declines in these numbers have spurred much of the recent study (as you'll soon find out).

Now, if I asked you to think of a seabird, I can almost guarantee you will not think of a black guillemot. Herring gull most probably, perhaps even an albatross or a puffin if you're feeling rogue? But a black guillemot (*Cepphus grylle*) is unlikely to come to mind. I would go so far as to say that we don't give many seabirds much more than a passing thought, despite the fact they are possibly the most accessible link we have between land and sea.

A member of the auk[23] family, the black guillemot is a hardy, duck-sized bird found on coasts across the northern hemisphere – an inshore breeder with a penchant for a rocky island and a harbour wall. Like most seabirds, they are long-lived (up to 29 years old) and invest heavily in laying

[22]The Royal Society for the Protection of Birds, established in 1889 by Emily Williamson, who sought to eradicate the trend of using feathers in fashion. With more than one million members, the RSPB is Europe's largest nature conservation charity.

[23]Pronounced 'awk' – it includes birds like razorbills, guillemots and puffins.

one to two eggs each year. Black guillemots in Europe and the rest of the world are not endangered, but populations in the UK and Ireland are at the southern end of their range and arguably aren't as secure. Over recent decades, a fall in numbers has awarded black guillemots a place on the UK's amber list of 'Birds of Conservation Concern', meaning care is needed to ensure they don't get upgraded to the dreaded red list.

In the spirit of general divisiveness, I believe these birds out-rival a puffin any day. Mottled, grey-white winter plumage moults into a summer spent in style: a chocolatey-black and snow-white uniform with killer crimson legs, slender head and a blood-red gape to match. The gape describes the inside of a bird's open mouth. Honestly, the black guillemot's is *fabulous.* During summer, these little guillemots look quite literally suited and booted at all times, thanks to their oval, white wing patches, which may remind you of a killer whale's false eyes. Nature's doodles repeat like that sometimes.

Across the Scottish Isles, black guillemots are affectionately named 'tysties'. Norse words like this still casually feature in Highland and Island conversation – especially around nature – most often nodding to a Norseman's intangible relationship to seabirds. I must say, it took me a good while to realise that 'tysties' were not some sort of local haggis-related delicacy. (Of course, I didn't like to be a bother and just ask someone to explain.)

North Ronaldsay is the northernmost island of the Orkneys. Famous for its migratory birdlife, world-class bird observatory and seaweed-eating sheep, North Ronaldsay lies on a similar latitude to Karmøy in Norway and is well on the way to the Shetlands and the Fair Isle. It was here I hoped to spend time with the colony of tysties that revel in this tiny island's rocky shore. For them, North Ronaldsay appears to be the gift that keeps on giving.

Both tysties and this island were new to me, and I was eager to meet them.

'Yeah, she's not normally *this* rough at this time of year!' a crew member admitted to a rather ashen-looking family. He shifted his internal ballast to counter the listing of the ferry as if policing a dinner queue at a 30-degree angle was completely normal. 'But, this is nothing!' he added with a cheerful wink as the dad made a swift exit to the toilets. The waves were vast. I had set up a basecamp of sorts in the restaurant at the stern, watching a fading Scottish mainland and a raging wake boiling through the window. I imagined mariners scattered across the North Sea, keenly waiting for the forthcoming *Shipping Forecast*.

I think I like a rough sea. My brother Tom and I were lucky enough to have a Mirror dinghy with a wooden mast when we were younger. She had all the proper rigging, a butter-yellow hull and red sails. We rinsed every corner of the Exe Estuary in her for years. Mum named her *L'oiseau*, meaning 'the bird' in French. I thought the name ridiculous at the time – but it stuck. And surveying a furious, darkening North Sea, with gulls, gannets and fulmars helming the surf as elegantly as a pianist does the keys, I realised *L'oiseau* was the perfect name. Because out there? That is a bird's domain. To mark this revelation, I bought two mini bottles of wine. One down, and my inner soundtrack rose to a glorious crescendo, so I headed out on deck to join my fellow rosy, windswept passengers.

A soup of turquoise waves pummelled the hull of the ferry. Immense offshore wind turbines rose from the churn like a stone circle commemorating an impending *War of the Worlds*. The sun appeared for a beat and flashed gannets brilliant white. In truth, I felt very enlightened and a bit Scotch mist (apt Cockney rhyming slang) as I finished mini bottle number two, following the contrails of birds as they danced across their North Sea like paper planes. I assumed the confidence of the solo traveller and became that person who wanders up to couples and families, offering to take their photos.

Docking in the capital city of Kirkwall by midnight, I discovered that during the crossing, my bike had acquired a comrade in the form of a laden touring bike. Its rider was an older man who had left the British mainland in search of change. Unlike my meticulous itinerary, time had no place in this man's journey. His only certainty was his bunk for that evening, which turned out to be at the same destination: a hostel 2 miles away. My headlight had run out of battery, and so had his rear light, so we sensibly decided to become one long tandem: he at the front and me following. The night was thick and dark; the lights of Kirkwall glistened in the distance. St Magnus Cathedral was illuminated like a red sandstone beacon. Known as the 'Light in the North', it was founded in 1137 by Viking Earl Rognvald and is the only intact medieval cathedral in Scotland. The man and I didn't speak on our ride, and after finally making it to the hostel, we bid each other goodnight.

I had two days to kill before I continued to North Ronaldsay. For those isles not joined by relic causeways from Churchill's time, inter-island transport usually means a short flight via Logan Air. Therefore, my desire for low-carbon travel was met with confusion by the tourist board, who couldn't fathom why I would choose a three-hour cargo ferry over a 10-minute flight. In the end, I was the sole foot-passenger booked on the weekly supply ship that takes goods to North Ronaldsay every Saturday. Even so, based on the number of grey seals and gannets in the harbour at Kirkwall, I was looking forward to my private boat trip.

Over the next two days, I rode 100 miles around the Orkney mainland and some southern isles. I don't think I have ever ridden such incredible stretches of road. Sure, I lucked out with clear, sunny skies (worth noting that weather like this was *not* usual for August), but the absence of traffic, hills and potholes made for the best riding I have ever experienced. Swathes of cotton grass, heather, pollinators, birds and silence hugged miles of long, straight road. I barely realised how many miles I was chewing through until I collapsed with a heap of pizza in the evenings. The first day I rode to the mainland's west coast. Fuelled senseless by the smell of sea salt, I left an effigy of myself at Twatt Church. Please note that Twatt is a genuine parish. Apparently, its road signs keep getting stolen by tourists. (Can't imagine why.)

Making quick progress on my bike, I approached the coast at Marwick Head – home to Orkney mainland's biggest seabird colony. I heard it before I saw it. The RSPB uses this astonishing mass gathering at Marwick as an indicator of general seabird health, noting alarming and inconsistent gaps in species numbers since 2007. Like orchards, seabirds have good and bad years, governed by many variables. The year 2021 marked the end of the fourth national census. This Seabirds Count has been replicated at more than 10,000 sites across the UK and Ireland, providing vital insight into seabird abundance, productivity, survival and dietary changes for five years. We can infer a lot about the state of our climate overall simply by observing seabirds, and counts like these are urgent and essential.

Generally, most seabirds return to places like Orkney to spend a raucous summer – mating, rearing chicks and fattening up – before they endure an unthinkable winter spent feeding in the open ocean. I cannot stress how much we underestimate this phenomenal feat – it sounds horrendous. Someone once told me that seabirds are 'bastard hard', but maybe this isn't the case for tysties, which prefer a more home-based life.

Refuelling with snacks on a beach just along from Marwick Head, I watched shoreline kelp. Its colour was such an intense shade of copper; I imagined it was rusting before my eyes in the midday sun. A mini starling murmuration spilled across the bay – it felt busy, but not with people. Things were *happening* here. London's hustle seemed a lifetime ago. Seals were bottling, floating upright with their heads poking out between the waves – they reminded me of bald men in hot tubs. Arctic terns darted overhead as I pedalled on, blowing my mind that the bird with possibly the longest migration of any animal on Earth could be so nimble and tiny. Their annual pole-to-pole 30,000-kilometre flight pales my cycling endurance to a pathetic blip.

The road carried me and my bike easily, pulling us away from the coast for short sections before inevitably returning us to its shores. During these meanders, I noticed inshore wetlands and how their marshy mosaic creates a haven for waders. Vast flocks of oystercatchers were engaged in excitable chatter, stirring in a fluster as I whizzed past them. The mountains of Hoy rose in the distance. I wanted to climb them. The next day I cycled to the Bronze Age site Tomb of the Eagles on South Ronaldsay, where I ate two pasties while getting mobbed by Arctic terns. Orkney made me want to do everything – at least, it made me feel like I could.

I soon discovered these islands were ideal for two things: generating renewable energy and drying laundry. Wending my way around the astonishing Skara Brae, it became clear that life here had been sewn long ago, yet Orkney feels surprisingly with the times and ahead of them sometimes. Orkney is home to the highest concentration of domestic wind turbines in the UK – despite having one of the smallest populations of any county (around 22,400). In

2009, Orkney became the UK's first 'smart grid' and operated more grid-connected devices than any other site in the world. All this makes it a natural base for the European Marine Energy Centre. One in 12 households in Orkney operate on their own renewables, and it's not uncommon to see bed sheets and tea towels strung up on washing lines alongside a mini wind turbine.

Wind energy has become so productive here (especially in winter) that Orkney was a net energy exporter in 2013 and 2014 and has avoided producing an estimated 50,000 tonnes of carbon emissions to date. While switching to the natural resources offered by the immense winds and waves is no doubt a wise move, it's more complicated for renewable devices offshore. The pressure to switch to cleaner energy stems from the hefty oil and gas exploitation across the North Sea in the early 1970s. By 1985, the UK was the fifth-largest oil-producing country in the world. Toxic by-products from this drilling remain a significant threat to seabirds, choking, sinking or starving them. Oil coats feathers, preventing flight – this threat is well known across global seas. Rising temperatures, oil spills and a danger to sea life drive the argument for greener energy sources. Orkney leads the world procession in the adoption of renewables. As a nation, Scotland is generating close to 100 per cent of the electricity it needs from renewable resources, moving away from fossil fuels at a pace the rest of us cannot ignore. In the UK as a whole, 2020 was the greenest year on record for generating coal-free electricity. (Let's hope it wasn't an anomaly, eh?)

Unlike the turbines we've come to know across the landscape, tidal-stream devices are fully submerged underwater. They harvest energy from the kinetic flow of ocean currents around narrow channels, islands and headlands, converting it into electricity. It's a smart idea. Some of these devices yield astonishing electricity output.

The Pentland Firth separates the North Sea and the North Atlantic tidal streams, and the mad strength and speed of its tidal races (more than 30 kilometres per hour in places) have claimed many human lives. Danger to life aside, this firth remains one of the world's most productive sites for tidal energy. Since 2018, the Scottish energy sector has gained planning permission for nearly 400 tidal stream turbines, bolstering what aims to be Europe's largest tidal turbine array.

However, the more urgent matter for some scientists is understanding how these devices interact with species that share their environment. Much like the situation with the harbour porpoises off Strumble Head in Wales, we need to step back before worshipping the renewables narrative. As tysties are shallow divers and lovers of rich benthic environments (the seabed, the riverbed, *etc.*), they are in danger here.

Following a tumultuous history of being hunted in Icelandic waters, a ban on hunting tysties was only recently enforced in 2017. Similarly, bycatch in gillnets – often left unattended for days – is a pressing issue for all diving birds. The estimate of birds killed like this around the world each year hovers at around 400,000 currently. The first proper study of tysties was in Canada in the 1930s on Kent Island. Since then, monitoring technology has been slow to modernise. Research into the physical collision of seabirds with offshore turbines (both wind and tidal) is growing, as are investigations into how our little portly auks might fare in a sea that has composite blades whirring about in the current.

Dr Elizabeth Masden is a research fellow from the Environmental Research Institute at the University of the Highlands and Islands based in Inverness. I'm still furious at everyone for not telling me about this magical-sounding institution when I was choosing where to study. Anyway,

Elizabeth has spent the best part of the last decade filling gaping holes in tystie-related knowledge. Several people highly recommended her expertise to me. 'Tysties were a species that no one was studying in the UK,' Elizabeth told me over Zoom. 'There had been some work in the 1980s and 90s, but not much since. It struck me all of sudden that we might be about to put tidal turbines in areas where tysties feed and nest, so we really need to learn about them before that happens.'

When it comes to monopolising island tidal races in our bid to trump carbon, tysties seem to get in our way. (Or are we getting in tysties' way? Discuss.) The tystie and the turbine want the same slice of habitat, which puts tysties in a dilemma similar to the one faced by the harbour porpoise. It's a game of 'who gets there first?' with most of the points up for grabs loitering around Orkney and nearby Caithness. Research investigating the potential tystie-versus-turbine story has been slim thus far; however, Elizabeth co-authored a paper in 2019, writing of this emerging paradox surrounding the quest for green energy. Yes, renewal energy represents an essential step towards damning fossil fuels to the history books, but it could also herald a significant human-wildlife conflict. There appears to be an awkward overlap between MPAs and the areas with the leases that govern Scottish wave and tidal energy sites. Winner takes all?

Dr Daniel Johnston studied tystie foraging ecology for his PhD, supervised by Elizabeth. Together, they now work closely to probe these merging worlds. Our 'other' guillemot (*Uria aalge*) is bigger than a tystie. More common too, it is widespread around UK coasts and also feeds far offshore in winter. These guillemots are also among the deepest diving birds in the world, capable of hunting at 180 metres below the surface. Little is known about the diving behaviour of our tystie, however. Working on the hunch that tystie diving habitat coincides with the locations of tidal turbines, Elizabeth attached depth loggers to birds on the nearby uninhabited island Stroma. Tysties were recorded as diving to roughly 40 metres. 'Meaning, yes, they could coincide,' said Elizabeth simply. Other observational work on tysties during Daniel's PhD involved GPS tracking. Along with Elizabeth, he wanted to determine whether tysties visit the fastest-flowing water where the turbines are positioned. 'The diving ability was impressive for birds as small as these,' Elizabeth admitted. 'Although there is location overlap, the birds might not use the fastest-flowing water where the turbines are.'

A separate study in 2015 also used depth recorders and GPS loggers to study tystie diving behaviour in Northern Ireland colonies, finding a strong preference for daylight dives at less than 2 kilometres away from the colony. Most dives were U-shaped, but each tystie had its own technique. These researchers stressed the importance of knowing these things if we are to hope for a future that involves both tysties and tidal turbines. Direct collision with the blades and mortality of the birds remains a threat. Feeding locations, feeding behaviour, prey behaviour and the seabirds' horizontal underwater movement are just a few of the areas of research about which we are in dire need of enlightenment. The classic ecological fear of 'too little, too late' drives nearly all the growing research around tysties. 'We're still chipping away. There are so many unanswered questions,' Elizabeth warned.

Rory Crawford is the Bycatch Programme Manager for the BirdLife International Marine Programme. A Glaswegian with a hint of Viking, he spoke with me over Zoom. Rory loves a tystie and is the kind of guy you want to sink a few pints with while putting the world to rights. He also is an instrumental figure informing and crafting policy on seabird bycatch around the UK and Europe.

'For me, tysties feel like a totem, a species that has been there since the start of my career … it's a personal one.' He described the energy developments around Orkney as 'the conundrum we can't avoid at the moment. The industrialisation of the sea is at a scale we've never seen before.' Like Elizabeth, he cautioned about the grey areas. 'We're in this zone of not entirely knowing what's going on, but should we be tackling climate change in a way that threatens the things that make the world worth living in?' Good point. Whatever we do, there remains much questing to do. And tysties for me to see. To the north!

A curlew dropped a sloppy turd next to me on the pier in Kirkwall. Early and already warm, I queued alongside cargo, ready to board the ferry to North Ronaldsay. I got chatting to a crew member who brokered conversation with a random anecdote about a charm of goldfinches he once saw in his garden. I'm not sure what vibe I was giving off to indicate that I would take an interest in this, but he seemed buoyed that I was excited for him. I discovered satisfaction in announcing to someone that I was simply on a journey 'heading north!' without elaborating further because, in this part of the world, heading north is a feat in itself. The ambiguous potential of doing so surely spurred those first settlers to Orkney from mainland Britain, don't you think?

Stretching my legs along a wooden bench on deck, I felt greedy in my solitary pursuit as we set off. We made swift progress. Hundreds of seabirds cruised a towering cliff, assuming similar choreography to the dandelion seeds of May. As Kirkwall shrank behind us, the spire of St Magnus stood rigid like a rose thorn. Skies quickened. Seas darkened. I lost count of the lion's mane jellyfish that lit the way to the island and liked the way that no two were ever the same.

As I leaned over the side of the ferry, the breeze whipped my hair, bleached by the sun. I was tanned, strong and felt good. Well, of course, you would, approaching a white strip of sand surrounded by waters that looked like the actual Caribbean despite being further north than Oslo. For the next few days, I was determined to become one of the people here. Already clocking three seals sidling up to the pier, I pretended I was coming home.

Known as 'Orcadians', the people living on the Orkney archipelago are among the best people you will ever meet (like the Welsh!). A proud people, they are kind, measured and loyal to their shores. A bit like tysties, perhaps? When Orcadians talk of 'the mainland', they mean the mainland of Orkney – not Scotland. They call Scotland 'Scotland', emphasising a healthy distance between the two. Tripping up on the gangway and flinging my water bottle across the stones as I fell, it soon became clear that the news of a visitor from England had circulated the island as fast as Covid-19. I was greeted cheerfully by a lady and automatically directed to the bird observatory where I was staying. Granted, I was the first visitor to spend more than a day on the island since the pandemic had begun, and I brought with me the potential to infect the entire island with the disease. All 72 residents had remained sheltered from it thus far, and at the time of my trip, tests for anyone travelling for work were still tricky to come by. So the thought of me being Visitor #1 since the pandemic was thrilling (but also terrifying).

In any other year, I would have been just one of hundreds of annual visitors to the North Ronaldsay Bird Observatory. Established in 1987 by Alison Duncan and located on what is often the first stretch of land birds see after setting off from the Arctic, the observatory is a bit of a Mecca for birders worldwide. Eager to see the birds that make landfall here during their North Sea expeditions,

North Ronaldsay is on every birders' bucket list. Hanging out at 'The Obs' (local speak), blowing away the cobwebs after working in 'the field' and sponging up decades worth of ecological records in the cosy bar with a hot tea and a mutton sandwich constitutes a day well spent.

Now I must admit that I am as likely to call myself a 'birder' as I am the Emperor of China. I don't own binoculars. I'm impatient. I never understood the hype when a brown booby blew on to British shores. I was hiking in Cornwall at that moment in 2019. Tens of people had driven hundreds of miles – some had even flown from Ireland – to try and see a brown bird that had been blown off course during its random wander from its regular haunt of Mexico. I am undoubtedly very naive, but I worry that the infamous, often aggressive 'life list' competition between serious twitchers risks losing sight of the species itself. However naive I am, I know I am not alone in worrying whether this behaviour deters some potential recruits to the cause.

Nevertheless, we can learn a lot from these enthusiasts and owe much to their patience and fortitude in pursuing rare feathers. Indeed, many birds have had improved fortunes, thanks to the devotion of birder to the scope. And over the following days on North Ronaldsay, I caught a glimpse of what drives this obsession, a glimpse of how seeing a bird that has permeated your thoughts for weeks can swiftly transform you into its doting apostle.

I was the only guest staying at The Obs and was greeted warmly by Darrell and Laura, who not only look after guests but are also residents and bird-ringers[24] in their

[24]Those who are licensed to attach small, numbered rings to the legs of wild birds to gather data.

own right. I soon discovered that the people here are an effortless blend of work, nature and community, and nearly all those I met who worked at The Obs were around my age. (I nearly keeled over with the novelty.) They were a healthy mix of those born and raised, those marooned during the pandemic on a bird-related placement, and those who had come here for a birding season then didn't have the heart to leave, fearing a piece of them would always be left behind. On a stunning August day, it's easy to see why. Later, Darrell quietly mentioned with a sly grin that there is a bit of a trend at The Obs for winter guests to be 'women escaping their husbands'. Noted.

As well as birds, the sheep across the island are world-renowned, being both quirky and the last examples of native British sheep. A rare, hardy breed reminiscent of goats, they live on the beaches and eat seaweed. Before I left home, a colleague advised that I eat the mutton out of respect for this fundamental livelihood. It was so delish. The annual North Ronaldsay Sheep Festival is a grand occasion. All the islanders and their extended families on the mainland flock together to champion their unique culture through art, music and the maintenance of the famous Sheep Dyke, which runs 19 kilometres around the island perimeter. The devotion to community here is pure.

Dumping my things in a room with a sea view, I was itching to walk. The island is tiny – its area just under 7 square kilometres – and apart from the airfield strip, pretty much everywhere is accessible on foot. Following a dusty track across fields of dandelion and escorted by some sort of pipit, I climbed over the stone dyke for a full-frontal of the North Sea shores. I'll be straight with you. There is no ambiguity here, no tentative build-up and purging of my emotional blockages when I see them because I saw the tysties – loads of them.

Swathes of rocky slabs lay before me. Clambering down, I selected one that looked like Pride Rock and lay on my stomach like Simba in *The Lion King*. I thought of Stephenson's quote from the train toilet a few days before. Yes, the journey to get here was fun. I loved the variety of it. But begging to differ, Stephenson, mate – this destination is something else. The bird that I had thought about for so many weeks was finally mincing about before my eyes. I watched them for a couple of hours, feeling soporific in the sun.

Tysties nest in pairs and live in social groups of less than 50. You may be pleased to hear that there is no physical distinction between males and females, making one feel less inept when even trained scientists struggle to tell them apart. Nests are usually identified by having splatters of (very on-brand) red poo around the entrance. As I found on the mainland, the shoreline was a happening place. My arrival was met with flashes of black and white erupting into hilarious shrieks, like some primary school PE lesson. Some tysties flew out to settle on the water among gulls and oystercatchers. Others darted into rock crevices like the one below me. Next to shearwaters and fulmars, tysties look strangely camouflaged and are challenging to see. This congregation of hardy survivors had been flung together by the North Sea soup. When bobbing on the sea, tysties often gaze to the horizon before racing back to shore, congregating in noisy little groups.

Such vocalisations are iconic to tysties, captivating ecologists for years. When I returned home, I caught up with Elizabeth Masden's former PhD student, Dr Daniel Johnston, who is now a research ecologist for the British Trust for Ornithology (BTO). After completing his PhD viva, Daniel felt suitably compelled to recycle an old Halloween costume, dressing up as a cardboard tystie and charging into the sea. An act that automatically validates all of his thoughts. (The video is still on Twitter – can recommend.)

'Tysties are really intelligent birds,' he told me, smiling thoughtfully over our video call. 'And they're unlike the other auks. They stick around. They're this *survivor*. And they have a sweet song, which not all seabirds have,' he laughed. This nest song behaviour is crucial to a tystie's courtship routine and was first studied in detail by Sten Asbirk in 1979 on some Danish islands. Although I cannot tell whether I saw moments of romantic merit on North Ronaldsay, courtship can be seen throughout the breeding season. Its disarming comedy aided, of course, by the fact that tysties on land look like waddling butlers and seem to be capable of displaying a rainbow of emotions in record time. Safe to say that I detected regular subtext chuckles from Asbirk while reading his write-up.

Most seabirds breed in remarkable synchrony, continuing to puzzle science. But studies suggest that the bundle of tystie behaviours I'm about to describe have evolved, extending the breeding season to match the availability of food – an asynchrony ensuring that everything has a fair chance. The ritual begins with a bit of 'staccato piping' to fend off the curious (*i.e.* me). Asbirk informs us that this may mature into a 'veritable piping concert'. A flash of raunchy red gape is expected here. I must say that the first time I saw this gape in real life, I instantly thought of a pair of Christian Louboutin heels I once lusted after for 5 minutes, the underside of which had the classic red lacquer.

Shortly after the concert follows an erection of their necks (usually in males), with gape half-open at all times, tail cocked, breast puffed while mincing about on their outlandish red legs. A bit of amiable scratching and aggression towards a future mate won't go amiss, either. If in water, no matter, because a bit of parallel swimming, perhaps a flirty little game of 'it' or underwater 'leapfrog' with a female, can be an equal turn-on. As is 'bill-dipping' and 'twitter-waggling', both of which go a long way in a male securing a second date. If the female is interested, she'll let him know by lying flat, neck and bill outstretched, and tail cocked. The male will then arise and 'trample' her back for some time. After the interlude, she then 'raises and throws the male off'. Sounds about right.

Asbirk describes this all as 'so infectious' that the excitement might plague the rest of the colony, initiating the same behaviours in 15–20 other pairings at any one time – meaning that this carefully choreographed, often patient, performance results in a sexual carousel. Ought we take note? Because from a grossly provocative exposé of ankles during a rendition on the fortepiano in the 1800s to a series of late-night 'likes' sweeping the 2017 social media archive, it's fair to say that the art of human courtship has somewhat … evolved.

Watching tysties muck about on the rocks in front of me, in between erratic flights to a bright blue sea, I could understand Elizabeth's love for this bird. 'They just have *so* much character,' she said, grinning over the screen. 'It's fascinating to watch and try to understand. They're incredibly social, which explains their quirky behaviours.' When a pair breed successfully (wahey!), the female tystie incubates the egg for up to 40 days. Like many seabirds, they co-parent, are known to defend the nest and engage in equal feeding of their one or two chicks. So very wholesome.

I still find it funny that a bird as performative as the black guillemot would economise so intensely on their nest. There must be a trade-off in there somewhere as, for many birds, nest design is critical in attracting a suitable mate. Elizabeth agreed. 'Yeah, it's not a "nest" as you'd expect. It's just some rocks.' She described how she and Daniel would be walking across Stroma and encounter an eider duck nest, 'the most luxurious nest in the world! And then we'd find a tystie nest, which is essentially two rocks that have touched, and an egg sat in the crevice.' Ask any ecologist who has studied tysties, and it turns out these crevices are the thorn in their cargo-trousered side. They are one of the main reasons tysties have been notoriously challenging to study. Remaining loyal to nest locations, Elizabeth described how tysties would return to nests annually and that at each site, she would often find the same bird, suggesting a tystie's loyalty to the doorstep.

Rory Crawford also explained that tysties 'like their nests to have an entrance and an exit'. Civilised, yet #humble. I like it. 'And they're the classic "once bitten, twice shy", so when they've been caught once, and you want to retrieve a tracking device back off them again, they're like, "Argh, *you're* the tube who caught me before – nice try!"'. I shook as the sonic boom of his laugh came through my headphones. Tysties are naturally cautious birds and tend to creep to the back of the rock crevice, especially when a group of researchers want a cheeky probe in the name of data. Elizabeth, too, reeled off hilarious anecdotes of dangling upside-down to reach her hands into deep holes while tystie chicks aimed squirts of poo at said hands. 'But, weirdly, because of all their caution,' she said, 'you feel a huge amount of respect for these birds themselves and how they monopolise these tight spaces – they're really, really smart.'

The following day I walked the island's perimeter – just under 20 kilometres. A few of the bird-ringers from The Obs had invited me to tag along with them while they finished ringing fulmar chicks. 'Any tysties?' had become their general greeting to me. I felt like I was being initiated into some sort of society. Following the track that ran alongside the sheep dyke, we hiked clockwise. The winds were easterly, and the ringers said these conditions were perfect for migrants drifting south from Scandinavia. Conversation was effortless and mostly bird-related. We asked each other what we hoped to see on the walk, as though pre-drinking before a wild night out. Alex, the chattiest of the group, told me that if puffins are 'pop music', tysties are 'indie'.' Exactly. Our edgier auk.

As we walked on, I saw that impossibly fluffy fulmar chicks occupied nearly every other crevice in the dyke – clearly a vital place of refuge. What followed was an entertaining episode from *Chunder from Fulmar:* a series. One of the girls sensibly tied her hair back before squatting down by the fulmar to attempt to ring it. Immediately, she was met with incessant squawking, clattering bills and bright-orange vomit, which exploded onto the ringer's hands and clothes. Fulmars' natural defence mechanism, this orange acid gave rise to their name, 'fulmar', translating to 'foul gull' in Norse. These chicks were at their awkward, greasy stage (we all have one), fidgeting within an orb of downy feathers. But soon, they would moult into the sleek, seductive plumage of their parents.

Despite ringing tystie chicks earlier in the season, the guys gallantly re-enacted for me the infamous palaver of prostrating themselves across the rocks to access a feisty tystie chick. Rain or shine, there have been daily surveys like this for more than 30 years, and we have this brilliant bunch to thank for much of the knowledge of seabird movement up here.

Enormous rocky slabs fringed the entire west side of the island. Tysties came and went in disorderly flurries, deftly navigating the chinks and chasms that housed their young. When they fly into land, their red feet are splayed, and their wings are madly propelling. When they move on land, there's a hint of penguin waddle about them. I imagined that watching them underwater would be a total game-changer, though – 'streamlined' being the foregone conclusion. I watched as tysties landed on the water with a jaunty splash the way a human-atop-rubber-ring might exit a water slide. I couldn't get enough of them.

Calm waters defined the day, but the high winter seas the locals told me about sounded unsettling. It's no secret that more storms are the sidekick to climate change for the entire planet. And, worryingly, thanks to higher concentrations of carbon dioxide in the atmosphere, recent research has measured clouds across the planet absorbing 7 per cent more water than before. Higher cloud saturation essentially intensifies a feedback loop between rising temperatures and unstable weather. What a to-do! Our centrally heated, jet-setting, soya-latte lives are paving the way for more numerous and extreme storms.

As shallow divers, studies have shown that tysties are at risk of death and injury from being hurled into the rocky shore when a turbulent sea batters them. Other research suggests these events can also flush benthic prey further out into deeper water, away from the assumed shallow comfort zone of the tystie. Speaking to Daniel, it makes sense that overwinter survival remains the most challenging time for them. He described 'wrecks' of starved birds seen in the wake of severe storms and explained that breeding success could also be hampered. 'We saw nests get flooded out by some horrendous storms,' he recalled of his time on Stroma during 2017. 'A few of our study nests failed. Coupled with a storm during that summer was a supermoon bringing huge spring tides. It was sad to witness.'

The Stroma nests that Daniel was studying in 2017 with Elizabeth are similar to those on North Ronaldsay: open, rocky and utterly exposed. 'If the storms increase in number and intensity with climate change, that is terrible news for the black guillemots,' Elizabeth sighed. She added how, because they nest on boulder shores, tysties expect a certain maximum tide height, so 'if a storm surges further up the shoreline, that directly affects the breeding success of these birds. The chicks and eggs just flush away.' But remember that tysties are cunning, and they may be able to anticipate dangerous environmental shifts. Daniel told me they were able to move their breeding habitat if they had to. So, yes, some North Ronaldsay storms filled tystie nests with sand. However, upon investigation, they discovered the birds had already relocated to a different part of the island before this happened. Do they know something we don't? Or are they just paying more attention than we are?

It seems a tystie's loyalty to its home shores could be both a blessing and a curse. A blessing because they are spared (what I imagine to be) hellish winters in the open ocean. However, if they remain on increasingly exposed, volatile shores and cannot move quickly enough to safer ground, we have a problem.

As well as the stormy threats they're facing, because they prefer to nest on the shoreline, tysties are also vulnerable to land predators. Calculating invasive species like cats, American minks and rats have long had a history of denting tystie populations. Arguably, they were the primary stressor that confined current tystie colonies to far-flung, sparsely habited islands devoid of mammalian predators. And I must tell you about something else that is unfolding as I write this in September 2021. Surreal numbers of guillemots and razorbills are dying. These seabirds are washing up emaciated and en masse on eastern shores from Orkney to Norfolk. Although wrecks are relatively typical following winter storms, it is incredibly concerning when they occur in September. Kayakers, swimmers and fishermen have also reported bizarrely close encounters with confused and devastatingly disorientated guillemots swimming in the shallows along our east coasts. I reached out to Rory over email, desperate for answers and genuinely terrified that this was the start of a new normal. There was no denying how alarmed he sounded. He told me that while exact causes of this phenomenon are still unknown, scientists are scrambling to determine the role of extreme weather events generating toxic algal blooms in seabird feeding hotspots around the UK. When seabirds hunt in these waters, they ingest lethal toxins that seize their bodies, slowly killing them. Some studies predict that climate change will increase incidences of these algal blooms in the North Sea. It's just one thing after another, isn't it? Does anyone else's head hurt?

I asked Rory if seabirds have pulled the short straw in general. 'We're always caught in this guessing game with seabirds. Dealing with what *might* happen,' he told me. 'Like if there is increased turbidity, and kelp beds get ripped up, then that's absolutely a problem for the tystie.' Kelp beds are an interesting one, actually. The image of towering golden fronds and dancing sea lions in underwater shafts of exotic, foreign sunlight comes to mind. But in the British Isles, kelp beds are (like seagrass, of course) a contender for our Ace of Spades.

We have the most diverse range of kelp species in Europe, spread across nearly twice the area of our terrestrial woodland. Tysties and kelp occupy the same habitat a lot of the time, along with many worms, crustaceans, lumpsuckers and such. Now tysties enjoy a variety of prey (and are lucky to do so), but they have a long-standing relationship with butterfish – a small, 'eel-like' thing found along the UK's rocky shoreline – and butterfish love a kelp bed. Daniel has observed a strong association between tysties and kelp in the field, but when we spoke, he mused it is 'likely we could lose these habitats to increased winter storms'.

Kelp is usually anchored to the seabed by a 'holdfast' (an anchor-like root system), but potent storms can uproot it like a carpet. Kelp and seaweed are also highly temperature-sensitive. Already around the British Isles, the temperature of some of our waters has risen by one degree Celsius within the last 40 years. But before you begin to picture summers spent frolicking in British waters without experiencing circulation difficulties, remember how warmer, more carbon-saturated waters become corrosive, breaking down the calcium compounds that cement these habitats together. Remembering seagrass, we've learned that ocean acidification knows no bounds. What follows could be a desertion of ecology. Cue the toppling dominos. (Again.)

Kelp beds function as a vital corridor that connects land and sea via seabirds. 'Many seabirds, not just tysties, bring nutrients back from the sea that otherwise wouldn't make it onto the land,' Rory continued. Seabird guano (poo) in the tropics, for instance, 'feeds' coral reefs, the resulting productivity of which influences marine systems across oceans. 'The thing is,' he added, 'seabirds are "the flow" between the terrestrial and the aquatic. There's something special about that. They connect these environments like no other animal.'

I can honestly say my walk that day was one of the best of my life. I had left the fulmar ringing team to their vital, vomity task and roamed on alone. Back at home, south-west England grappled with record temperatures and heaving beaches, but North Ronaldsay was spared. I'm still unsure what I'd done to deserve having pristine white sands and turquoise seas all to myself. With every passing minute, I fell more deeply for the island, and I will forever associate it with the pop-rock band, The 1975, as their music was on repeat in my headphones for most of that day. Feeling as 'indie' as a tystie, I picked my way across rocky shores, peering into tidal pools and abandoned crofts, shepherding friendly sheep, madly waving at farmers, and straining my eyes towards a fuzzy Shetland in the distance beyond the lighthouse (Scotland's oldest and most intact!). I adopted the slow, measured walking pace of someone for whom time has no bearing. Solitude had a different flavour here. I felt lucky. Freer than ever. Passive abandon.

But I wasn't alone. A huge rush of wind made me instinctively throw both hands up over my head. 'Bonxies' – great skuas. The guys at The Obs had warned me to wear a hat to protect me from a (not unheard of) scalping, but I had 'forgotten' in favour of a tan. Quickly ramming a cap over my head, I stole a glance at my antagonist. Dark, large and a bona fide 'bad arse' – as a friend of mine would say – it banked ahead of me, ready for a second go. Skuas – both great and the more dainty Arctic – are in their element on Shetland and Orkney. Agile and clever, they are notorious for their tactical 'mobbing' behaviour to protect their nest or even steal prey from others. Arctic terns do it, too, but they are not nearly so intimidating as a skua.

With Orkney and Shetland both suffering population losses of more than 80 per cent since 1986, the BTO classes the Arctic skua as the UK's fastest-declining seabird. Caustic and daring in their pursuit of food, a skua's methods appeal to my dark side. Rotund little birds (like the tystie and the puffin) are an ideal target for a bit of food piracy, or 'kleptoparasitism'. One reason for this is their sloppy table manners. After catching long, skinny fish like sand eels, tysties and puffins don't bother to conceal them. Instead, on the return journey to the nest, sand eels dangle from their mouths like floppy cigars for all to see. Carrying sand eels crossways like this creates easy pickings for a nimble robber that can snatch a fish without even making contact with its unlucky carrier.

As a proud left-hander, one of the things I can't stop thinking about is the fact that scientists are wondering whether tysties show 'handedness'. Some adult tysties and puffins carry prey like sand eels with the head always on the same side of the bird's bill. I noticed, too, that those tysties that carried food on the surface of the water were hesitant, often panic diving. This reaction seemed much quicker for them than taking to the air to evade ambush. Often huddling in a group, tysties may also fly in from the sea together in a bid to confuse and disarm a lingering food pirate. A paper in 1986 found that skuas conducted the majority of attacks on tysties, and persistent pursuit can result in the tystie dropping or abandoning its prey altogether. Oh, mate. Elizabeth told me, 'You root for them! You sit there shouting, "GO ON! You can make it! *Get in!*" and then a skua steals it. You know the skua needs the fish too, but you can't help rooting for a black guillemot.'

Contemplating piracy on an island like North Ronaldsay was enough to send me into a *Treasure Island* spin. Parasitism of tysties, in particular, is frequently observed by Daniel in his fieldwork: 'There's often a whole chain of piracy: skuas would steal some food, then a black-headed gull would have a go, then a herring gull would snatch it.' I wondered whether this is a growing problem for tysties, but it seems the pressure is nothing new. The same 1986 study found that, although skuas benefited considerably from stealing tystie food, tystie breeding success was unaffected unless the rate of attack was unreasonably high.

'It's more that their overall *diet* is crucial for us to observe,' Daniel said. 'Because their responses to tidal turbines, storms, temperature changes, parasitism – it all comes down to prey' – and whether prey is resilient. We've seen the danger in showing favouritism in a prey item. The terminal decline in UK Arctic skuas and kittiwakes, for instance, is owed mainly to a terrifying slump in sand eel numbers. It is 'LAST ORDERS!' on repeat for these birds. Dreadful times.

Sand eels feed on cold-water plankton, and more than 90 years' worth of ocean data (thanks to that Continuous

Plankton Recorder!) has directly linked warming sea surface temperatures to its rapid decline. Put simply, this reduction in food for sand eels has meant they are not as numerous themselves or of sufficiently fatty quality for the seabirds that rely on them. Have skuas resorted to robbery in desperation? It's hard to say, but there is no doubt that skuas would be on an even more slippery slope if the tysties and their dangling prey weren't around to target.

Now tysties have been known to enjoy a sand eel, along with butterfish and other treats. Both Daniel and Elizabeth recalled a certain tystie on Stroma 'constantly bringing back sand eels … its chicks were *so* fat and healthy!' Daniel laughed. Feeding niches between individuals in a colony can range from kelp beds to sandy areas, tidal streams and rocky shores. Such individual variation in prey choice and foraging habitats seems typical of tysties. 'It's a mixed picture,' Elizabeth admitted, 'but not as bad as it would be if they only ate sand eels.' A bonus, perhaps, to have a flexible palate? 'Yes, but they've been shown to favour prey, and if that prey has had a bad year, then it's not going to be good for them either,' she said.

What we want to avoid is change at the *colony* level – that's what will trigger the alarm system. We have to hope this eccentricity within a black guillemot colony – the varied eaters and trademark diving behaviours – is sufficiently robust to keep tystie numbers afloat. We want this bird to ride out the storms to come.

Eventually, I circled back to The Obs and stayed (or rather, laid) a while on South Bay. In my daze of sore feet and dehydration, what I thought were mounds of intertidal rocks suddenly materialised into snoozing, snorting, flatulent mammals. Easily mistaken for heatwave survivors on any British beach, a bob of grey seals (what a collective noun!) was hauled out on the sand, swivelling their dog-like heads in unison towards me. Whiskers twitched. Grinning and keenly aware they were assessing me, I tiptoed around the back of them.

Only bird prints accompanied mine on the sand, for the tide had long washed away any other signatories. Chilled, turquoise waters soothed my swollen feet. Before long, five, seven, ten – wait – *twelve* enormous, whiskered heads were in the shallows keeping pace with me. Inquisitive and unafraid. It felt like these seals had been spared as participants in the Great Conditioning Experiment ™, in which nature has learned that humans are an immediate threat. They just wanted to know what, where and who I was.

Buoyant in my happiest mood, I assumed this whole performance was a personal invitation to go swimming. The seals didn't seem to enjoy the whole undressing situation, flicking away at me with tail splashes and such. But in my white pants and sports bra (big mistake), I waded up to my shoulders and waited. And they came. The clarity of the water offered a porthole to their underwater playground. In a short space of time, their repeated probes into my world grew less bashful. I'm not sure how long I was in there for, but I stayed until it became too cold to bear, feeling elated, freezing, branded. The seals and I exchanged a transaction for which I never want a refund.

That evening I ate alone in the dining area, thinking as I soaked in the panoramic view of the camping meadows and sea beyond. A commotion of voices filled the reception. Before long, I was spontaneously bundled into the back of a van with Darrell, Laura, Heather (daughter of Alison, who founded The Obs) and a few others. 'Fancy a swim in a tidal pool before the sun sets?' they asked me. Hath Her Majesty a crown?

On the short drive, Heather stopped the truck, turned the engine off and engaged in friendly chat with every person we passed. *Actual* chat. Far more than obliging small-talk in a queue. A rare thing. I admired it immensely.

The sun hung low over the horizon, a golden medallion against a pink, lilac and baby-blue mackerel sky. Others were already at the pool that was left behind as the tide retreated. Beyond them, the setting sun silhouetted an enormous bull seal above the swell. Westray and the other islands rested in a distant blue blur to our left. Easy laughter, honesty and #acceptance was the general theme of the evening. Feeling inspired, it came to me that moments like these are what life might be about, and the simplicity of it stunned me – a group of people, a van, an island and Pot the dog.

Being my last evening here, I wanted to finish where I had started. Salty and wind burnt, I later dangled my legs over the harbour wall. Around thirty tysties busied about, dancing day into night. 'They have their own little thing going on,' one of the bird-ringers had said to me, 'but only those who *know* them love them.'

The sitting tysties looked like they were resting on a red whoopie cushion, their legs folded underneath their tuxedo. Despite the dusk, the clarity of the water meant I could watch the seals from South Bay swim over – just checking in. Arctic terns wheeled above. This time, the tystie song was more of a whistle, like a childless swing in the breeze.

It turns out this scene of tysties on the harbour wall is in many ways the key to their survival because it links them to people. Elizabeth, Daniel and Rory all spoke of tysties' immense adaptability to nest in human-made places. Nesting on walls, houses and under balconies allows tysties to act as a 'gateway' species to all seabirds. 'People may feel more of a connection with a butterfly or a bee,' Rory had said. 'But seabirds are telling us the story of the climate in an immediate way – humans just stay a bit too far away from them to hear what they're saying.' A solitary black guillemot on my right was gazing out towards the mainland. Time to go. I had some literal shit to deal with back in the south of England. Coming?

CHAPTER SIX

Dung Beetles

September was on the move. Summer drifted into its finale of stable weather before autumn ripened. At home, the sea was at its warmest, and I swam almost daily during those weeks. Then, like some sort of imposter car journalist, I suddenly found myself behind the wheel of an electric car outside Tiverton Parkway train station waiting to pick up my friend, Gina. You see, having become a rail connoisseur of late, I had grown bored of trains. Besides, I cannot lie – I really *do* like driving. (Sorry, Greta.) After all, a combustion engine, wheels and a progressive mind have lured us down the open road since 1892, when 20-year-old Frederick Bremer built the first four-wheeled petrol car in his home town of Walthamstow, London. Ah, those were the days, what! Anyway, here in the twenty-first century, I was keen to stay loyal to low-carbon solutions but was also motivated by a desire for timetable control.

Besides, I wanted to test whether an electric car could rise to the hype. I felt as qualified in this endeavour as Jeremy Clarkson might at reporting the Harrogate Bridal Show (though I would give generously to see such journalism unfold).

My friend, Nicky, generously loaned me her BMW i3 – a small, fully electric high hatchback made of carbon fibre and aluminium. On the ugly side for a BMW, I'll admit; even so, I was driving a BMW, and thus I blossomed into a certified *lad*. What Highway Code?

The i3 has certainly made waves in the electric vehicle market. I'm told its design is largely responsible for making electric cars more mainstream. Amazingly, sales of plug-in vehicles registered in the UK soared by 66 per cent in 2020 compared with the previous year, inching towards the government's 10-point plan (note: 'plan') to ban new petrol and diesel car sales in 2030. The thought is that by 2050 we might (*might*) be carbon neutral.

The battery of an i3 is 50 per cent bigger than when it was first in production, delivering a 33-kilowatt-hour surge of energy to the wheels. The environmental credentials of electric cars are debatable. The life cycle of lithium-ion batteries is not exactly low-carbon. This soft, silvery alkali metal powers our laptops, phones and electric vehicles (EVs), and lithium carbonate is even used as a psychiatric medicine to treat schizophrenia and bipolar disorder. An intensive mining process extracts it from the salt flats of South America, home to more than half of the global supply. Estimates suggest that nearly half a million gallons of water are needed to bring 1 tonne of lithium to the surface.

Recycling lithium batteries is difficult and expensive. Disposing of them can leak toxins into the environment. Widescale infrastructure to support this process isn't there yet, and in the UK, we park a quarter of our cars on the streets, which takes the spark out of charging an EV easily.

The carbon fibre body structure of the i3 (while lightweight and helpful for battery life) uses about 14 times as much energy in its production as steel, the more traditional car body. The race to electrify all vehicles is by no means an easy win. But then again, what is?

My destination in the borrowed i3 was Knepp Castle Estate. You may have heard of it – West Sussex's pioneering hinterland. Sitting alongside North Ronaldsay on a birder's bucket list, no doubt – and with good reason. Since 2001, more than 3,500 acres (1,416 hectares) have undergone a steady, astonishing transformation from decades of intensive farming to a redefined wilderness under the Gatwick flight path. Fences abdicated their reign so grazers could roam year-round. Wild deer, semi-feral ponies, and de-domesticated cattle and pigs were reunited with a bygone, wood pasture landscape. Relict floodplains were found and restored like an old painting. Water returned, writing a new story across a land that had shackled its course for too long. And then the wildlife came. Peregrine falcons, nightingales, turtle doves, purple emperor butterflies, long-eared bats (hello again!) and dung beetles. Some of these species we thought would never come back.

This is 'rewilding' or 'ecological restoration'. Rewilding is often misunderstood as the process of returning a landscape to its raw, natural form. Unfortunately, humans have already toyed so much with the land that the past is out of reach. In fact, both these phrases describe the process of allowing nature to have more of a say in how the land answers to change within the boundaries that remain.

Perhaps that all sounds a bit limp and fanciful? A time-travelling landscape pocked with large herbivores and flights of rare butterflies in place of combine harvesters? The trade-off there might seem potentially problematic. But there is no denying that 'rewilding' is a bit of a buzzword these days, even accounting for the word's many, shall we say, interpretations. Knepp is a triumph and shows us what could be if we let go. It is a reminder that positive change can exist within a human lifetime if we want.

'Rewilding is all about creating novel ecosystems and allowing nature to respond,' Isabella ('Issy') Tree told me once. Issy and her husband, conservationist Charlie Burrell, are the landowners at Knepp. She documents its inspirational journey of change in her remarkable book, *Wilding*. Friends, if you haven't read it – you must.

Naturally, I was visiting Knepp because I had read they had an impressive record of dung and associated beetles. Yes, dung – as in animal poo, excrement, droppings, etc. A couple of years before, in 2018, a master's student counted a staggering 11,633 dung beetles on Knepp over five days. To those so inclined, such faecal real estate is not just impressive: it's enviable. Strange, I thought, to gather such numbers when 25 per cent of dung beetle species across the UK are rare nationally, with most lumped into the 'endangered', 'vulnerable' or 'near-threatened' categories. It seemed I had entered into a bit of a dung-shaped dilemma. So off to Knepp I went in my sartorial electric ride, hoping to discover the secret of its dung-beetle success.

I hadn't the faintest bloody idea what I was doing. Having only recently grasped how to 'Pay at Pump' for petrol at my local supermarket, the combination of figuring out how to plug in the car to charge it and maintain sufficient charge en route to my destination had a strong chance of bruising my mood. Ironic that someone so reliant on plug-in devices would feel intimidated by a plug-in car. I'm a good driver, but I felt suddenly inept after only 3 miles in the i3. Until then, I had only ever driven a manual car, my dad's beloved silver Ford Focus, and I was fooled by the minimalism that comes with having a battery in place of an engine. My left hand grappled for a gear stick that wasn't there. My left foot followed suit, plunging involuntarily for a clutch that also wasn't there. And as I lifted my foot off the accelerator, I kept worrying the battery had drained because, suddenly, the car would brake noticeably, often slowing to a standstill when approaching a junction or traffic light.

Most EVs have regenerative braking, a sustainable way to slow down and conserve battery life. I'm interested. Some say regenerative braking is so effective that you can cruise London's streets in an i3 without ever touching the brake pedal. Once you get used to these quirks, you are in for a bit of a wheeze. For a start, EVs are eerily quiet. Potentially deadly for cyclists but good for viewing unsuspecting birds. Next, these rides are *brisk*. The i3 packs a particular punch, being one of the few EVs to send electricity to its rear wheels instead of the front wheels. Safe to say, I became well versed in shaming some diesel and petrol cars, pulling away from roundabouts and green lights rather spectacularly. (Between you and me, I *may* have morphed into an archetypal BMW driver in those moments.)

I mentioned I was travelling with a friend this time. Like me, Gina enjoys having a lovely time outside, and she is exceptionally talented at spotting wildlife and telling people about it. So, I thought she would be good to have on hand to alert me to all the things I wasn't noticing when we got to Knepp. Being around Gina also has me on the edge of a laugh at all times, whether we're sourcing 2 a.m. chips in Bristol or navigating deepest darkest Cornwall looking for visiting whales and trying to educate the world about their migration on social media (before furiously deleting all evidence when we realised we were filming a group of rocks). Gina and I have always joked that we are like brother and sister, though which of us is the brother in this relationship remains up for discussion.

A long road lay ahead of us, but we managed three hours driving to Petersfinger near Salisbury before the battery needed a boost. St Peter (and his finger) may be enamoured with the fact that there were numerous 'pod points' to charge the car here, and we found our first in a Morrisons car park. My uncle had recommended an app called Zap-Map that highlights the nearest charging points along your route, ranging in price and length of charging from 'rapid' to 'slow'. New to the app, I was slow to clock on to the various options, and we were in that car park for more than an hour, inadvertently having chosen the slowest. At least charging turned out to be easy, requiring no more than the well-practised manoeuvre of plugging cable into socket. It was reasonably priced, too, with nearly 80 per cent of the battery restored for less than a fiver. Sure, an electric car like this might set you back more than £30,000, but there is some satisfaction to be found in not losing £60 every other week to a petrol mogul.

The car's realistic range was about 140 miles, and toying with the numbers became the game of the journey. Much like fuel levels, battery life is influenced by all sorts of things in the car: air conditioning, the radio, heating, satnav. You can watch the miles pile back on or drain, depending on what the driver and passenger are demanding at any given moment. Somewhat synonymous with how fossil fuels are consumed, to be honest. However, unlike fuel, you can increase your battery life while on the move if you wish. For example, after a second top-up beyond Salisbury, the car's display showed we only had enough charge left to travel 73 miles, yet Knepp was still another 94 miles away. Thus began a fervent appeal to the theatre of battery-saving that only the most committed of us hone over the years.

Little did I know I had been training for this moment for the past decade. Every ploy I had ever tried to prolong a rechargeable device gave me strength. Game very much *on.* And much to Gina's frustration on a muggy September afternoon, I turned off the air conditioning, satnav and radio. Steadfast in the slow lane, we maintained a modest 63 miles per hour. I barely touched the brake as though it were a bed of coals. Silent pleas were sent to the electricity overlords, asking us to be spared another hour in a supermarket car park. *Get me to the beetles.*

We peeled off a gridlocked M27 at last. A27 flowed into A24 and branched into tributaries of smaller, quieter roads. Woodpecker Lane. Swallow Lane. Roadworks, traffic and charging had tallied our total driving time to more than 8 hours. Yes, sure, we could have flown to Dubai within that time to promote our new line of #vegan lip tint. But we preferred playing restless children attempting a low-carbon journey to wonderland who really needed the toilet (and desperately wanted to see some dung).

I've been mulling for some time how to sweet talk dung to you. I am truly curious as to your immediate thoughts. Are you repulsed? Has a bad experience prompted lifelong aversion? Or are you simply indifferent? However you might feel about dung, I'm pretty sure we like beetles. A friend of mine once said she sees them as 'indestructible', which suggests a level of admiration, I think. Even so, when it comes to understanding their actual function, it's easier to move on, isn't it? Yet, it occurs to me that dung, in all its forms, offers us a rare opportunity for lifelong education. Indeed, we've learned that our grey long-eared bat prefers a more flamboyant approach to defecation. I can't say the same myself, but I did enter a regrettable phase for about a year:

Parent: 'How was school?'

Me: 'Poo.'

'Much homework tonight?'

'Poo.'

'What time is Rachel coming over?'

'Poo.'

This most favourite retort was maddening for everyone else and hilarious for eight-year-old me. Aside from the obvious, a good many synonyms describe this most fundamental of biological happenings, some of which I will experiment with throughout this chapter: egesta, discharge, ordure (sounds like an ordeal), ejecta, turd (she will *never* admit it, but this is my mum's favourite!) and, my new favourite, ejectamenta (some sort of private school Latin motto?). Anyway, the fact is, the world wouldn't exist without it, and animal droppings buttress our countryside like little else.

Predictably, we've been swift to recognise dung's value for our society. Dried dung has long been bricks and mortar for many communities around the world. Across India, dung is still culturally regarded as some sort of universal blessing. Where trees are sporadic, people burn dung as fuel. Our crops have been fertilised with dung since farming dawned more than 8,000 years ago. Yet, it seems that a sort of global dung divergence has emerged. Some cultures continue to harvest fresh dung for worthy, medicinal purposes. Oklahoma holds an annual World Cow Chip Throwing Contest at which contestants compete to throw dried chunks of cowpat. The record stands at 81.1 metres, unbeaten since 1981.

Cows, of course, augment our lives in other ways. Milk? Useful! Ice cream? Here for it. Cheese? Stupid question. But here's the thing – I would confidently bet money on the fact that you have never, ever concerned yourself with another by-product of a dairy cow. Nor have you worried for the UK's dwindling dung, let alone lamented the average consistency, health or content of it in a standard cattle field. Well, I'm with you. Until recently, I rarely thought about dung either, unless it worked its way onto my shoe. Overlooking dung is just the way things have panned out over the years, I guess. But the best dung is rapidly becoming the stuff of legend. And it's high time we sorted that shit out and allowed ourselves to be schooled by the humble pat and the life – yes, *life* – within it.

UK dung beetles are part of a beetle 'superfamily' (and it is quite super!) called Scarabaeoidea, and within this superfamily are two smaller families: Geotrupidae and Scarabaeidae. It's easy to feel intimidated by the sheer number of vowels in these names; never mind try to get your head around the beetles they describe. The main point to flag is that these families comprise of 'tunnellers', 'rollers' and 'dwellers' (I imagine most human families could assign each of those labels to at least one of their members). You may recall scenes from nature documentaries that show Tiny Sturdy Beetle™ rolling an enormous ball of dung across the savannah? That visual won't serve you well in the British Isles, I'm afraid. Dor and minotaur[25] beetles in the UK sometimes push rabbit pellets about wantonly, but with nowhere near the same panache as beetles in other climes. But let's not hold that against them.

Dung beetles, of course, have an appetite for the brown stuff. Like it or not, they quite literally live for it. Yes, other beetle families feed on dung, but about 60 species of 'true' dung beetles live in the UK. And it's these species that 'truly' eat the dung itself. As you'll come to realise, lots of other beetles – like predatory rove beetles – live in the dung pat itself, looking to feast on fly maggots, soft-bodied invertebrates, fungi and other beetles. But these are 'dung *loving*' – as opposed to having a true palate for it. Beetles themselves are the most diverse group of insects, with more than 450,000 species alive across the planet today. That's 25 per

[25]Yes, half-man half-bull in Greek mythology, but also an *actual* dung beetle and a certified bad arse. Relatively common in the UK, it enjoys sandy grassland and heathland. Look it up – it really is something else.

cent of animal life so far described. Beetle bodies have three sections: head, thorax and abdomen, encased by an external ('exo') skeleton. Language reveals more. In Latin, beetles are called '*Coleoptera*'. And in Greek, *coleoptera* translates to 'sheath wing'. It turns out the beetle's unique selling point is its double-wing situation. Rigid, armour-like wings, known as elytra or wing-cases, on the outside protect the delicate flight wings underneath, much like a sword rests in its scabbard. This tells us that the true defence for many beetles is, in fact, a quick, aerial getaway.

Most animals contribute to ecosystem function, but dung beetles are the Amazon Prime of this functionality. Resolute in their mission and disarming in their efficiency, they would probably take over the world if they could. Bees and pollination are an obvious force across the countryside. Somehow, though, dung beetles remain forever in their shadow, despite the fact they often occupy the same environment and offer competitive ecosystem services. Dung beetles are the (travel-sized) operations managers of the ground we stand on.

Continuing this brief detour into an office environment and dung beetles would be the fast talkers, the multitaskers – ultra-responsive, reactive, contributing noteworthy points in a board meeting while simultaneously texting, tweeting and ordering flowers for their mum. Not necessarily sociable, but the sort to wear a beret to post-work drinks and feel good about it – such assuredness they exert in their role here on Earth. As we'll find out, dung beetles trigger a bewildering division of labour among thousands of other species, all playing a part in removing a pat from the field and restoring the integrity of the soil. Soil that, in the UK, stores more than 10 billion tonnes of carbon. Soil at risk of total fertility loss over the next 30 years, the degradation of which already costs England and Wales alone more than £1 billion every year.

Arriving at Knepp shortly after 5 p.m., we staggered stiff and slightly dazed into reception like a couple of occupational hazards. I had arranged to meet with Penny Green, the ecologist for the estate (a storybook waiting to be written, surely?) for a quick whistle-stop tour of the beetles and turds likely to be on offer for the next 24 hours. Mainly for aesthetic reasons, I hoped to see a violet dor beetle. Dung beetles are around throughout the year, but violet dor beetles are often seen more towards autumn. Five years ago, Penny discovered one during a survey of Knepp, a stunning, iridescent hulk of a thing. Violet dor beetles hadn't been recorded in Sussex for more than 50 years. While making tea for us in her office, she described how, over the 15 years of landscape regeneration here, the return of browsing mammals and woodland pasture had produced enough organic dung of every shape, size, texture and scent to support an abundance of beetles. 'Organic dung is missing across so much of our landscape,' explained Penny. 'In most sites, dung beetle populations are naturally low because they've been surviving on anything they can find. But if you rewild – and increase dung by having livestock out and about and not kept in barns – give it 10 years or so, and the beetles will be self-sustaining.'

Even from the car park, Gina and I sensed we were somewhere special. Not just because we had heard about Knepp and its delights for years, but because there were large yellow signs around the site informing us of 'free-roaming animals'. (And if you don't perform an internal cartwheel at that prospect, then you need to go sort yourself out and come back to me when you're ready.) The roar of the road felt close, but noises that we had never heard before were coming from bushes opposite the car. Utterly strange, yet familiar all at once. Trundling our things in a wheelbarrow to the camping ground and catching our hair in low-hanging branches, we happily entered our 24-hour escape room.

A short walk from the car park nestled in a meadow framed by ancient woodland, Knepp's campsite is quite the hype – for campers and glampers alike. Choosing a sunny corner pitch by the rain-water shower block, we faffed with sleeping bags and fly-sheets, soon prioritising dinner in a nearby pub. On our walk into Dial Post village, the light had been golden and low, revealing the most enormous red deer stag we had ever seen, relaxing in a copse, almost within touching distance to our left. Stopping dead, we gripped each other's forearms magnetised. 'I wouldn't be surprised if rainbows started coming out of his arse, to be honest,' murmured Gina.

Later, on the walk back to the campsite, I was so full of food and red wine I half wondered whether I might be with twins. The night was bright and, although distant cities ensured the sky wasn't totally dark, a single shooting star streaked overhead. Prepping our tents for a windy night, it took me a minute to realise the background grunting I could hear was not from the neighbouring tent but rather from the adjacent field where a couple of Tamworth pigs ruffled the pasture's edge. In *Wilding*, Isabella Tree explains how wild boar roughed up the edges of pasture like this in ye olde England, overturning soil and stimulating new growth across the understorey. Hunted to extinction in the seventeenth century, wild boars haven't played a role in our countryside for many years.

The 1976 Dangerous Wild Animals Act banned the release of wild boar onto any estate – Knepp included. But the team at Knepp noticed they could replicate a wild boar's influence in a less threatening form. So Tamworth pigs – ginger, large and loveable – were recruited as the understudy. (And very successfully.) Hailing from Staffordshire and a rare breed themselves, Tamworth pigs have an affinity for oak woodland, snuffling acorns and snorting roots, earthworms and everything in between. Tilling the soil like a traditional plough, Knepp's pigs open doors for species that haven't danced at this party for centuries. One of Knepp's most celebrated achievements is the rebound of the purple emperor butterfly – an outlandish, showy thing that had no place here before the pigs came. But, thanks to said pigs and their ability to encourage the growth of larval food plants like sallow, the butterflies returned, and Knepp is now home to the UK's largest breeding population of purple emperor butterflies. Can't argue with that.

Waking early to the sound of more pigs and even the odd owl – a tawny, I think – was idyllic obviously – the stuff of stories. I wondered whether, once upon a time, the whole of England woke to such an LP. Planes formed an orderly aerial queue into Gatwick. Quickly realising our drastic oversight in the cooking department, Gina and I stole into the campsite's forbidden 'glamping kitchen' on the hunt for caffeine and some healing baked goods. I would have died for a croissant. But, alas, none were to be found.

Further drastic oversight confirmed that many other nature seekers also visit Knepp, and they were also up with the dawn pig chorus in feverish lust for said caffeine and healing baked goods. Soon the kitchen featured a small group of sleepy waifs and strays, swapping milk splashes and spoons of instant coffee along with our hopes and dreams for a weekend in the wild. The banter was niche and grossly nerdy, everything that is inaccessible about conservation in a way. But most of these people exert a special kind of warmth that is a challenge to dislike in person.

'To the dinosaurs!' shouted Gina, stepping forth through an enormous steel gate as we entered the Southern Block – the wildest of the three areas that divide up the estate around the roads. A violent clash of red deer stags locking antlers met us immediately. Testosterone simmered on the brink of boil, ready for the rut. A something-spotted woodpecker sought refuge in a nearby oak, the roots of which massaged a ground that looked untidy, chapped even, but not unpleasant, especially when a nuthatch bounced around the folds of the root. (By the way, if you ever find yourself vexed applying liquid eyeliner, Google a nuthatch and thank me later.) Deadwood relaxed into pasture, around which skirted a last breath of ragwort. Hedgerows were strewn with sloes. Gina darted around with her camera like a sheepdog rounding the flock. Everything just felt *bigger*. And I liked it. I really, really liked it.

It turns out that not many people dedicate time to the pursuit of dung and associated beetles. Somehow, prodding poo for answers is not aspirational. But, luckily for us, Darren Mann exists. Nobody does dung quite like Darren. He works for the Oxford University Museum of Natural History, and together with his wife, Ceri, they run the National Recording Scheme for dung beetles in the UK. He offers management advice to landowners in a bid to work land in favour of dung beetles as a remedy to improve landscape health. I caught up with him over Zoom. His office was classically academic, books upon books. Chaotic but meticulous. 'I'm one of the few academics who can admit that they do shit research – I literally spend all of my time looking in shit. You've got this whole ecosystem just in this one piece of crap.' He recounted tales of bug hunting through the Hindu Kush mountain range in northern Pakistan, travelling by yak during a blizzard to a shepherd's hut and burning yak dung for warmth. 'Smoky, but not unpleasant,' he added.

While on furlough during the pandemic, instead of nurturing sourdough starters, Darren and Ceri took their campervan around the UK, filling gaps in dung beetle records and building the national map. They have even rediscovered species that ecologists thought were extinct. Although based in the UK, Darren travels the world, piecing together what makes dung so vital. He's the kind of person who recalls his experience with elephant poo as 'like being a kid in a sweet shop' or who enthusiastically claims there is 'nothing better than breaking open a piece of dung in your hand … you never know what you'll find'. He reels off scientific names of beetles as lyrically as a waiter might relate the contents of the specials board to you at a fancy restaurant. Hectic diaries pushed my call with Darren to after my trip to Knepp, but crikey, it was worth it. Over Zoom, I was teleported back into the lecture hall. It was probably the best lecture of my life. 'Most people don't realise the UK even has proper dung beetles. It's not even in their periphery. But,' an almost wicked laugh, 'we would know if they *weren't* here – there would be shit everywhere.' According to Darren, people just get bored with beetles. I agree, but I think it's because they don't know them.

Time to get into the basics. A grassroots approach will do, so we'll start with the pat itself. The average dairy cow individually produces up to 15 tonnes of the stuff per year. Figures from 2018 estimate nearly 10 million cattle roam the UK's pastures. That's a lot of ordure – around 80 million tonnes. But there's a lot *in* a pat too. Essentially, it's a dump of undigested plant mulch, bacteria, nitrogen, phosphate and sulphur – all of which the soil would like back at some point. In her office, prior to our dung foray, Penny described with wonderful hand gestures how a healthy pat has rings like an Audi logo. Three to four concentric circles. 'It has structure: peaks, craters and a good crust. Rather than a great big runny mess that we are normalising across the countryside.' One could interpret the quality of a chocolate brownie with the same criteria. Surely the best have a good crust but remain gooey in the middle? Well, it's the same with a good pile of mammal excrement.

Healthy pats are good for science, too, as Darren explained how a single pat could give you information on the soil type, age of dung and location in England simply by analysing the species that colonise it. 'It's why insects are so useful in forensics,' he remarked, 'they have a very specific biology that can be traced.'

You may have noticed that Earth is regulated by patterns or cycles. The four temperate seasons, day and night, sleep and wake and eyebrow thickness are a predictable few. Dung decomposition is another. The whole process doesn't take long, but it must happen. And it's one of the few cycles that we can follow right through to completion if we so wish. Naturally, on a dung foray, you'll encounter pats of different ages. There's a sweet spot of time for beetles to monopolise a pat, ranging from seconds to a couple of days. Charlie Burrell of Knepp Estate is notorious for lying alongside dung, 'straight from the rectum!', and watching how dor and other beetles assemble within seconds. 'The great, glorious trifecta!' exclaimed Darren – praising the cow-dung-insect love triangle.

I must mention some elements of dung beetle hardware. Beetles are among the most successful animals on Earth. In the cast of dung beetles, this is in part because evolution has garnished them with the shovels, helmets and sensory equipment needed to negotiate Turd Mountain. Since they must also have sex on the pat itself, they need various adornments to engage in battle for a mate. At this point, you may feel overwhelmed by the imagery. But, like the beetle itself, we soldier on.

Beetles here range from 2–68 millimetres long, and their decor includes spikes, rhino-like horns, knuckle-duster ridges and more. Head-to-head combat between dung beetles often occurs inside a pat, with all these prongs and protrusions ensuring quite the conflict. Observe a close up of a particularly ornate male, and they honestly look like a walking Swiss Army knife: nail file, blade, letter-opener, toothpick. If a crisis occurs and you have a dung beetle to hand, I'm pretty sure you'd be covered.

Claiming a fresh dump requires serious planning from these insects. Distinctive antennal 'clubs' – tiny feathery fronds rammed to the seams with sensory cells – are helpful. Effectively, this is a dung beetle's nose. Sensory organs often cover the body of a dung beetle but are also in abundance in the antennae. What's hilarious is that while most insects use these kinds of appendages to seduce and sniff out erotic hormones of a potential mate, dung beetles use them to sniff out faecal particles until their eyesight can lead them to it. Because it is but nature's will that a successful shag directly depends on the availability of good dung. The pat is where the magic happens, and the mere smell of it thrusts a dung beetle into utter rapture. And here is why dung beetles are having such a rubbish time across the UK: quality dung that supports this essential dynamic is decreasing. What dung is left is rarely organic, and the rest is often in the wrong place at the wrong time. Instead of roaming in pasture, cattle are moving indoors in barns during the winter. Such disruption to the precious dung cycle leaves dung beetles short of food and a mate – sad times.

Once a pat is ambushed, the moon landing begins, which can happen quickly. Some ecologists observed up to 50 beetles arriving by land and air in less than 20 minutes of a cow having a dump. What a superb fleet of horny dung-miners! 'These beetles are usually the first to arrive at a pat and initiate breakdown, along with dung-feeding flies,' Darren beamed. He explained how the first beetles to arrive ferry teams of bacteria and fungal spores from the last pat they visited and inoculate the new dung with early successional microbes. You can think of this as analogous to eating kefir yoghurt to improve your gut health – it's often a good idea. A small army of predatory mites no bigger than a millimetre (and oddly similar looking to sesame seeds) hitch a ride on these beetles too – 'like an Uber', Darren shrugged. Once at the fresh pat, they leap off and feast on the larvae of flies buried within. 'Often, these flies are pests to farmers, so having this predator-prey dynamic within the dung is vital.'

Some dung beetles in the *Onthophagus* branch of the scarab beetle superfamily are so ornate that they look like a cross between a gladiator and a tank. One of the UK's remaining six species in this group is even the same hue as the British Army's combat uniform – olive-green and field-brown flecks. Its head assumes the shape of a spade, presumably for shovelling dung about. The legs of tunnelling dung beetles remarkably resemble trowels, and they assist in deep excavation. And it's this process of working through the dung – to feed, mate and breed – that aerates it.

'If dung is left on its own, it undergoes anaerobic respiration of microbes, which does not end well,' admitted Darren, explaining how this leads to a concept known as 'pasture fouling', where dung is left unchanged on the pasture. Not only does this lock up nutrients, but it also kills the grass below. Bad news for crops and livestock alike. 'But get beetles in there doing their thing, and this buries the dung in the soil, opening it up to other insects that will continue to decompose it.' This 'bioturbation' adds nitrogen and phosphates to the soil, enriching its composition and increasing grass growth. Once the dung is in the soil, rove beetles (dung lovers) arrive, feasting on fly larvae and eggs. And then the worms come. They, too, smell the new dung, taking sections the beetles have perforated and tugging it further down on a deeper journey into the soil.

These successional waves of invertebrates mean that, in theory, one field with a few healthy cattle could constitute an entire archipelago of life – something that other species have learned to monopolise at Knepp. Before I embarked on my own dung survey, Penny had commented how a single pat with its city of insects could attract jackdaws, long-eared bats, horseshoe bats, owls and many other species, all of which want, and need, a slice of the action. Even the white storks – reborn in a landmark project within Knepp after hundreds of years and now free flying – feed directly on the dung beetles themselves. Beetles offer all of this rent-free. What noble and helpful tenants.

Predictably, Darren's zest has sharpened the curiosity of many. Mary-Emma Hermand is an entomology postgraduate from Reading. She and I are of similar age and jostled similar meanderings in our early 20s. Conversation flowed easily between us. She described a nine-month placement with Darren at the Oxford University Museum of Natural History during her Zoology degree as the best year of her university life. 'I remember being the only one in my group at uni who genuinely enjoyed the insect stuff,' she admitted during our call. 'But I found a kindred spirit in Darren and could nerd out about beetles with him and not feel weird about it, you know? It's easy to bond over the overlooked.' Mary-Emma's postgraduate research has helped expose the chasm in our understanding of dung beetles in the British Isles. She focuses on 'cryptic' species: two or more species that look almost identical with the naked eye but are genetically different.

Aphodius fimetarius is a native European dung beetle first described by a guy called Linnaeus[26] in 1758. But, in

[26] Carl Linnaeus was a Swedish naturalist who coined the binomial method by which plants and animals are classified, ranked and named across the natural world. The 'Father of Taxonomy'. Bit of a legend, tbh.

2001, a genetic study discovered that its population actually comprised two discrete species that looked unbelievably similar. Mary-Emma sought to quantify what exactly the differences were between both *A. fimetarius* and *A. pedellus*. Using Geographic Information System (GIS) mapping techniques and Met Office data, an overlap emerged. Mary-Emma shared her screen with me during the call to reveal a map of the UK. Little dots showed *A. fimetarius* was less tolerant of colder areas and preferred sandy soils, even up into the Outer Hebrides. In contrast, *A. pedellus* was found in areas 2°C colder.

An unconfirmed hunch from Darren and Mary-Emma showed that these distributions align with the Gulf Stream, which brings warmth to the UK and north-west Europe. In 2020, the Gulf Stream was at its weakest for more than a thousand years, which is likely connected to climate instability. 'It all seems to be linked with temperature – our main interest now is the impact of climate change within a very important group of beetles. I ended up with more questions than I had answers!' Mary-Emma confessed. 'You can lose biodiversity without ever knowing it was there. And that scares me. That *really* scares me.'

What's more studied – yet still in desperate need of further investigation – is the role of dung beetles in the carbon cycle. Climate/carbon cycle feedback is a vital area of research to help us meet Paris Agreement targets (*i.e.* limiting global temperature rise to 1.5°C). Rebecca Varney is a PhD student working in a team from the University of Exeter. She led a study in 2020 that mathematically modelled the turnover time of carbon in the soil under different temperature scenarios. The paper's intensity of equations and formulae made me feel a bit sick, so I gave her a quick call instead. 'The uncertainty is just *huge*,' she told me. 'We're working rapidly to reduce that given how vital soil is in the fight against climate change.'

Vital indeed. According to the United Nations, the global population of livestock surpasses the transportation industry in the emission of greenhouse gases. A third of global methane, a gas more toxic than carbon dioxide, comes from the farts and dumps spread across pasture. You'd be forgiven for thinking beetles play some sort of role in locking that up in the ground during the bioturbation process, in the same way that seagrass does in our seas. Darren had assisted with a 2013 experiment in Finland whose findings suggested dung beetles can help to do this, and I asked Darren whether it was fair to call dung beetles an 'ally' in the fight against the climate crisis. 'That experiment was artificial,' he admitted. 'Nature is just so bloody complicated, so if you remove ninety per cent of life within a pat, *then* measure its carbon sequestration, the findings don't reflect reality.'

Darren continued by iterating the famine of research in this area. Nearly 10 years on, and only two or three studies have followed suit, he explained. In 2020, one of them concluded that dung beetles don't actively accelerate a notable reduction in emissions and simply cannot match the pace at which they are released. They can't bury dung fast enough to make a marked difference to emissions. 'We don't have enough data to make broad conclusions. But dung beetles burying dung and introducing microbes can only have a positive impact on greenhouse gas reduction. There's no other way around it.'

I'm almost impressed that people have developed a different way of thinking in a relatively short space of time. Essentially, the animals around Knepp are livestock:

cows, pigs and ponies. But these animals assume a completely different guise when they are permitted to *Be! Free!* throughout pasture. Darren described how along with landscape connectivity, we're losing an essential combination of dung types. My mind raced. Take deer, for example. Roaming the UK since the Ice Age, their smaller droppings prevent big beetles from breeding on them. In areas with deer (and their small poo), this tips the balance in favour of smaller beetle species. 'So, you get these extinction filters depending on what dung is available. If you have only cattle in the landscape, you'll oust smaller dung-feeding beetles. If you have only deer, you'll miss out on the big ones. You need a mosaic of dung to facilitate the whole system.'

And this is what is so hot about Knepp. The astonishing library of dung on offer makes you want to whisper, *Hey, this is what the countryside should look like …* 'We've got dung from longhorn cattle, three different deer species, Exmoor ponies, Tamworth pigs, foxes, badgers …' Penny reeled off the specials board. And as I listened to her, I thought about how there will be dung beetles that specialise in all of these types of ejectamenta (I just wanted to say it once!) Penny had described Darren as the 'Tasmanian Devil of Dung' and admitted it was incredible to watch him at work. It turns out achieving faecal dexterity requires a lot of skill.

Following a smiley overview of the dung dos and don'ts around Knepp, Penny had left us to it. Plodding about the Southern Block, Gina and I sought out the best of the pats with which to receive our belated baptism of dung. Long-tailed tits that looked like flying teaspoons darted about an echo chamber of birdsong. We caught only flashes of them between the leaves, which flickered in the morning sun. A longhorn scratched an itch on a piece of deadwood while a small group of fallow deer assembled like parents at the school gates before leaping into a muscle of copse. Gina nearly lost it when a male fox, tangerine and glossy, trotted past. 'Fuck *me* … this is *amazinggg,*' she whispered.

I made up for what I lacked in trowel-like appendages and spade-shaped head with latex gloves and a clueless demeanour. Penny gave us very little instruction other than to plunge right in, wrist deep, and carefully sift the pat to search for life inside. But she had kindly armed me with the necessary wherewithal to inhabit the pat mentally (probably more helpful). Searching for living beings in animal waste still seemed a bit mad, but there was no alternative to becoming engrossed. Its texture was satisfying to manipulate, and, crucially, the smell was actually *not* bad. What we so quickly label as 'foul' smelled *earthy*, nutritious. As I rummaged in the middle of the pat, yellow dung flies mated around its edges.

Below the crusts of many pats, I mapped tunnel entrances created by those first colonisers after the oh-holy-excretion event. Some tunnels were drilled directly below the surface, some to the side. It's within these tunnels that brood chambers exist and the larvae develop. Removing as small a section of the lid as possible, the grass and matter the longhorn had digested was easily discernible. The dung that coated my palms and (parts of) my thighs was generous but not unpleasant. I wasn't grossed out, more like struck dumb. Because throughout the pats I poked, tiny life forms appeared. Tiny C-shaped larvae rested in the folds, grey-white and already with defined heads and mouthparts. Our sesame-seed-shaped mites scurried the interior topography of Turd Mountain. A ground beetle marched within it like a freedom fighter. Things were happening here that I'd never seen or considered. I didn't know what I was looking at, but for a minute, as I held court with a cow pat, I forgot that this scene is rare.

Upon returning home the following week, like a good student, I painstakingly sorted through and sent photos of turds to Darren. His reply simply read:

> Hi. Larvae are one of the 'lesser dung beetles'. Most likely *Aphodius fossor*. Diurnal. Hard to be sure from pics.
>
> Best,
>
> D.

A late arrival to a fresh pat, *A. fossor* generally flies at dawn and dusk. But records show it can be active during the day, depending on the local temperature and humidity levels. Eggs are laid within the crust and spend the summer maturing into larvae, which is the stage at which I saw them. Darren explained how the larvae of the *A. fossor* species rely on the cattle dung for their survival. They can eat deer poo but cannot breed on it. 'They need cattle dung in open pasture. Else it's game over.'

Two of the dor beetle species at Knepp are nocturnal. It occurred to me that instead of plying myself with wine and fantasies of magical stags with Gina the previous night, I should have diligently surveyed Knepp's nocturnal quantity. Feeling a similar resolve build as I had felt at my harbour porpoise fail earlier in the summer, I decided I was OK with only seeing dung beetle larvae. I later concluded it a high honour and privilege to have witnessed an extraordinary species at the very start of their subterranean journey.

I asked Mary-Emma what it is about dung beetles that make them so forgettable. 'We just don't recognise their importance. They're overlooked because they're unseen.' Speaking about the horse fields near where she lives, she told me that 'dung is no longer a resource. People tidy it up. Get rid of it. Take it away. That's just what we do now.'

The funny thing about British dung is its satirical exposé of our bad habits. Shit happens, right? And with this in mind, I'm aware that we haven't yet talked about *why* dung beetles are in trouble. Yes, if there were more dung beetles, that could be good for our greenhouse gas situation, but most dung beetles are 'pretty screwed', according to Darren. Information on the remaining dung beetle species exists primarily thanks to the records that he and Ceri have built. And even those population datasets are somewhat lopsided because there's hardly anyone else out there bothering to count them.

The truth is we are experiencing a rather severe drop in our supply of, um, droppings. Supply is truly reliant on demand. Rich dung with peaks, troughs and crusts is increasingly hard to come by because, like all mess, we simply clear it up like good boys and girls. 'A regular part of intensive farming nowadays is locking cattle away in sheds during large parts of the year and treating dung as passive waste,' Darren said. 'But this blocks the vital outdoor cycles that dung facilitates.' In an inconsistent bid to improve grassland, we're neglecting the natural richness that comes with the unimproved. Progressively warm and wet winters are bringing cows inside. '[Grassland] is cut so early for silage that it never reaches maturity,' continued Darren. 'It's then so full of nitrogen that when the cows feed on it during the winter, they get diarrhoea. We're rapidly on our way to ruining the countryside.'

Many cattle are also so heavily medicated with anti-wormers and antibiotics that their faeces are a runny, foul-smelling mess, devoid of life anyway. This inhospitable dung is a world away from the invertebrate metropolis that inhabits a pat at Knepp. Much research, notably from the University of Bristol, has demonstrated the danger posed to any insect that processes this medicated dung. It's essentially insecticide. They're going to die. Introduce some heavy rains into the mix, and we've got a genuinely shitty, druggy mess entering our watercourses. No, thanks.

Remember, the poo itself is food for many species. In places like Knepp, where the animals do not have routine wormers prescribed to them and are free to roam all year round, the beetles have a reliable food source, and they honeypot the area. Elsewhere in the countryside, the prospect is much bleaker. Darren sighed: 'There's a worrying lack of food for beetles. When animals are active throughout the year, they need continuity. And as they have annual life cycles, if the next piece of suitable dung isn't within their flight range, they'll go extinct within the year. It only takes one bad season of dung for a whole beetle species to disappear.'

Intensive farming isn't exiting UK landscapes anytime soon; it'll probably increase. But it's in the best interest of our livestock if we prioritise dung beetle diversity. Figures from 2015 estimate that dung beetles might be saving the UK cattle industry £367 million per year in services like pest reduction, better quality grazing due to reduced pasture fouling, reduction in the need for wormers due to improved bovine gut health and a bonus boost in soil quality. I've heard of a vet in Lancashire who is a confessed dung beetle champion – 'but we must engage more vets in this conversation,' Darren insisted. 'Medication and wormers are so persistent in the soil that they impact the cattle as well as the dung beetles.' Reducing the quantity of wormers in the herd is not only better for farming's carbon footprint, but it's also a real money-saver. 'Many vets still habitually use a blanket approach, medicating the entire herd preventatively instead of identifying individual cows and treating them separately.' This 'just in case' method is dangerous, never mind bloody expensive. We don't take paracetamol just in case we might get a headache.

Darren is currently working with veterinary groups to demonstrate pasture-wide remedies – natural, herbal solutions to pests that really do work. Chicory is a natural player in a healthy pasture but is also an effective wormer for grazing cattle. Chicory's high tannins also work to reduce methane production in a cow's stomach. Handy. Darren described how 'mob-grazing techniques' move the animals around to allow pasture to rest and give dung beetles time to decompose the original dung. 'We must view it as a functioning system and interconnect our decision-making with what we think is best for the land at the time.'

Ah, *The System*. Always acknowledged with such reverence among conservationists as if we truly know what it means. One dictionary definition of a 'system' calls it 'an assemblage or combination of things or parts forming a complex or unitary whole' and then cites a railroad as an example. We need to view our agricultural landscape like a superorganism made up of many parts. It may help if you visualise it as an engine. That's fine. However you think about systems, dung removal, livestock carcass removal, beetle decline, diminishing soil fertility and cattle diarrhoea are all symptoms of system failure. With the current so-called 'system', essential nutrients will cease to stick around. And as well as putting cows in sheds and removing dung from fields, the broader disintegration of our landscape is catalysing wider regressions across dung beetle numbers as some grim domino topple. Once dung beetles disappear from an area, it's hard for them to return. Our engine's warning light is blinking furiously, but it's just another light in a world of lights. We've become accustomed to it, and because we kid ourselves that we're still moving forward, we convince ourselves we'll be OK. We couldn't be more lost.

I mentioned earlier about beetles' double-wing situation. Darren's wife Ceri studies how far dung beetles can fly. Scientists haven't the foggiest about the distances dung beetles can tolerate between pats. Still, Ceri's work will fill vital gaps in our information, directly informing how we manage our landscape connectivity. Thinking back to Knepp, it appears its complex habitat embroidery makes it much easier for a beetle to move about. And roaming livestock seems to be a critical stitch in the final design – the frequent dumps of Knepp's cows, pigs and deer are literal stepping stones for dung beetles in that area of Sussex.

But connectivity will take time. Because although Knepp is the best habitat these species have had for centuries, the land surrounding the estate is a combination of intensive agriculture, habitat fragmentation and newly proposed housing developments. 'Yes, Knepp is suitable, but it's an island,' Darren paused. 'And if there is nowhere for the beetles to come from, if they can't physically get to Knepp, then their recolonisation around the county is going to be maddeningly slow.'

Indulge me with one final peek into the future of farming, particularly in southern England. Some dung beetles are doing well due to the enormous increase in deer populations during the past 40 years. You may challenge the notion that there is a lack of mammals roaming across our pastures when some wild deer herds are expanding annually by 30 per cent, but it's that dung assortment issue again. We need a selection box of dung to yank its associated beetles away from extinction, yet if temperatures rise as predicted, the chances of that happening are slim.

Recent research at the University of Exeter warned farmers to brace for an all-systems-change before 2100: a world that might be a whopping 5°C warmer and 140 millimetres less rainy each year. This world will make the south of England inhospitable for livestock. Arable crops will replace pasture. Livestock will shift north. Parts of the south-east might suffer from drought so acutely that farmers will have to abandon land – a situation that is already challenging areas of Spain and the south of France. Temperate becomes Mediterranean – poof! Just like that.

More investigation is needed to establish the threat level here accurately, but what of our dung beetles? I posed this question to Mary-Emma. 'Insect ecology is a web. And it's one of the most delicate webs in the natural world. There is just so, so much that we don't know. But the problem is we're finding it out as we go along.' I wonder...is this apathy or disconnect?

Penny told me how the dung beetles at Knepp have acted like Berocca to the soil, curing it of its epic post-war, agro-chemical hangover. Key observations from the latest (2021) soil samples across Knepp have labelled 'soil compaction' and 'nutrient deficiency' as 'absent', with an average of 16 earthworms per cubic foot of soil. I fancy a framed cross-section of this happy scene on my bedroom wall.

It's no secret that the more life living within the soil, the more carbon is being attended to. Yet, generally speaking, our relationship status with soil is terrible – a long and messy breakup. The rains of winter 2019/2020 introduced two different kinds of mud to the market near me in Devon. Type one is textbook: rich, dark, fertile, carbon-rich; mud with beetles ploughing it. But type two is creeping into the vernacular: rammed with silt, nutrients and chemicals; mud that is injured topsoil; claggy, runny and sticky. Dead.

In 2015, the Committee on Climate Change stated that since 1850, we'd lost 84 per cent of topsoil in the UK. It's thinning by 2–3 centimetres per year. Figures from 2017 estimate that we lose 24 billion tonnes of the planet's fertile soil annually. Even microplastics have found their way into the remaining topsoil.

Are we seriously going to gloss over the death of the seeds, roots and tiny life that allows us to have a tomorrow? Even for a human, that level of arrogance is pretty astounding: a joke that will never land. Soil health is both our saviour and our handicap, and dung is a panacea to rescue our soil from its terminal diagnosis. To flog the analogy, it's the prescription we keep forgetting to fill, let alone repeat.

Hope is a fickle notion but must be welcomed with open arms, across our farmland especially. At times, 'regenerative agriculture' and 'nature-friendly farming' seem like contradictory expressions, veering towards propaganda. Yet these terms define the crucial, holistic steps needed to pull Britain's landscape through rehab and help shape it as a more resilient participant. It's time we issue a belated public apology to the countryside and finally make meaningful amends. Minimising the physical disturbance of the soil – for instance, rotating crop sowing and grazing – has the proven potential to iron out kinks of the past. If anything, being savvier about soil makes financial sense. A landmark study in 2021 from Cambridge University and the RSPB found that land accumulates more capital when left to nature than when given over to humans. But then food security, of course, is the next question. Sigh.

The proposed Environmental Land Management (ELMs)[27] government fund is a subsidy scheme for land managers that incentivises a collaborative, solution-based approach, issuing financial rewards for good practice. My friend and colleague, Chris Jones, a beloved and unforgettable figure in farming (and beaver reintroduction), often tells me how regenerative agriculture 'works its arse off' when it

[27] Part of the 2020 Agricultural Act laid out by DEFRA aiming to emphasise environmental benefit. Among the farmers I know, it's been warmly received.

comes to provisioning for wildlife and future-proofing the herd. He explains: 'If you keep cattle away from the pasture that they've shat all over and rotate them around little paddocks allowing long rest periods in between – around 60 days – they won't have the opportunity to pick up parasites because you break the parasite life cycle.' After dung beetles have initiated decomposition, the birds that visit to feed on insects break apart the pat and expose it to UV light, which helps kill bacteria and viruses that the cow could otherwise ingest. 'You get this completely virtuous cycle of life simply by managing the grazing animals. Worming then becomes a pointless expense.'

Since Chris ceased regular ploughing on his land, Woodland Valley Farm in Cornwall, the soil's organic matter has nearly doubled over the past decade and increased in depth by 1 centimetre. Based on earlier figures from the Committee on Climate Change, that centimetre of depth would have taken nearly a century to establish had farming practices remained as intensive. 'Farming was once an exercise in applied ecology, but it has become one of applied chemistry,' Chris wrote to me. I felt his frustration.

But as Darren Mann explained with a sigh (a lot of sighing recently!), 'it's unwise for farmers to be vilified. They know the land like nobody else'. If livestock is left outside over winter, regenerative practices are a winner for dung beetle survival. 'But everyone's tired of conservationists telling the world that it's screwed, aren't they? So, conservationists need to *show* people what to do.' Darren stressed the impact an individual can have if they buy locally and choose to buy their meat from someone that looked after the animal well. Amazingly, he and several others are working on getting accreditation for 'dung-beetle-friendly food' on consumer packaging. Of course, asking the average consumer to gloss over the word 'dung' on their packet of sirloin might be a sticking point, but it's a challenge we must rise to. Because, as we now know, a healthy cow is a healthy pat, a happy beetle, a delighted soil, and back to a healthy cow. Satisfying, isn't it?

If you're a pet owner, you ought to know that wormers for our dogs and cats can accumulate in the environment, too. But veterinary surgeries are an ideal place for a spot of idle environmental health banter, are they not? We should talk more and be bolder about asking people who can change the things we can't. In the spirit of imparting invertebrate wisdom, Darren is relentless. His public speaking gigs at beer festivals are often so memorable he's been stopped in the street by people who have attended them.

Technology has revolutionised the accessibility of nature and insects especially. An app called iRecord allows you to photograph any plant or animal – beetle, bee, fly (ideally the 'bum view' apparently) – upload it, and then clever people ID it for you before making the data available to the public and for further research. 'So many amazing discoveries have been made just because someone with a casual interest in beetles took a photo of it with their phone,' Darren smiled.

Mary-Emma Hermand, a young woman on the cusp of a brilliant scientific career – 'too many times I've held a beetle tray of different species, only to be told that these can't be found any more. But as young people with a drive for change, we can be a bridge between generations to stop this from happening.' Introducing the people with the knowledge to the people willing to act would spark quite the ensemble, no? After all we've put them through, it turns out that the beetles are calling us to arms. We've got work to do.

I've figured out why the concept of rewilding appeals to me so much: it's driven by unrelenting nostalgia.

Tragic, perhaps. With similar vigour as a Tamworth pig roughing up a field margin, I dig up the past at every available opportunity, often carefully replacing elements of the present with it – a skill I have perfected since I was 10 and my parents' marriage ended. I'm learning that acceptance is a work in progress, but it's a needless battle and is, honestly, very tiring. For somewhere like Knepp, the same attitude applies. It's a place that allows us to see with our own eyes that an environment released from a rigid plan and trusted to build upon its beginning is better than before. 'Stop planning – and start watching,' someone once told me.

Gina and I had explored nearly 20 kilometres around the Southern Block on foot. Exhausted and smelling of dung and autumn, we emerged feeling scratched and uncivilised, yet looser. The clouds darkened, and beech leaves framing a pond glittered in a quickening wind that buffeted some mallards, making them look drunk. We plaited trails between free-roaming animal droppings on our way back to the car. Scanning crusts for six-legged pioneers, I decided that I would like to see dung beetles hopscotch pats on a more regular basis, please. Later, when writing this chapter, some late-night analogy had me likening their fate to sitting beside a loved one on their deathbed and hurriedly trying to comb through a lifetime of archives in their precious remaining hours.

I also considered that the world that turns below the soil would continue to abort our good sense if we were not careful. After all, it is already true elsewhere. Underwater habitats (freshwater in particular) hide a world about which we are equally ignorant – anyway, enough of these idle musings. Back at darkening Knepp, I needed to charge up the i3 and head home. Time to scrub up, look sharp and go see a river about a fish.

CHAPTER SEVEN

Atlantic Salmon

Borrowing an oak leaf from a sapling on the verge, I wiped something brown from my trainer before letting the leaf fall onto the strip of grass growing up the lane. No doubt, dung and its beetle brigade were still marching about my mind. The wooded lane was quiet, the wind low, and the air had a youth to it that felt good. Perhaps no one else had been down this way for a while. I liked thinking that.

October had arrived. Our passage into winter. Summer's footprints lingered on the beech and oak ceiling. Ahead, woodland glowed amber. After hitching a ride from Exeter to Drewsteignton from a kind pal who was (apparently) 'heading that way anyway', I made my way down narrow lanes and paths to the banks of the River Teign. Changing into my wetsuit, I was hoping it would equip me with a daring, courageous mood in pursuit of something I had never seen before. Fidgeting awkwardly in my neoprene body condom, I waited for another friend, Joe.

It turns out that Devon's ribbons of rivers are a bit of a hotspot for Atlantic salmon. But even as the girl next door, so to speak, I had never seen them. Between you and me, until now, salmon (like most species that live underwater) had just sort of passed me by. Barely worth a second glance because *that thing over there* is more interesting. I've since felt ashamed about this but then realised it's never too late to head down to the river and do a bit of poking about. Old or young, a bit of freshwater investigation will only serve you well. It had been more than a month since my whirlwind of folly at Knepp. Winter was dropping hints of its ETA in the morning air. Time to brush my hair and curtsey for His Majesty, the 'King of Fish'.

From Dogmarsh Bridge, I glanced up at Castle Drogo – Britain's youngest castle. Standing tall at a peachy 110 years old, this was the last castle built in England. In 1974 it became the first twentieth-century offspring of the National Trust, and it remains one of England's most popular properties.

Castle Drogo commands the view above the gorge on the northern fringes of Dartmoor National Park. Dartmoor historically was (and remains) one of the wettest places in England, and such is the power of nature that damp can permeate even the most solid of structures. During Drogo's youth, maddening amounts of leakage and structural damage occurred to the entire castle. Mission Watertight involved a national supply of scaffolding and tarpaulin. The restoration ran to tens of millions of pounds before the castle was finally unveiled, fresh and sealed, in August 2019.

Drogo was ahead of its time, as the concept of generating electricity using hydroelectric turbines in the River Teign was proposed by engineering firm Gilbert Gilkes and Co. Ltd in 1916. Following a few decades of faffing around, the original 1929 turbines were officially reinstated in 2016 as part of the National Trust's Renewable Energy Investment Programme. Now, more than 50 per cent of Drogo's activities are driven by renewable energy. And I think the staff should have full permission to dine out on that chestnut as often as they can.

The Teign Gorge itself is spectacular. A classic steep-sided 'V' under a duvet of trees, framed by upland heath and oak of international importance. Deep within the gorge, an ancient woodland is returning. Much like the castle overseeing it, Fingle Woods is a national treasure. Think Dame Judi Dench or Gerri Halliwell in that Union Jack Dress. Again (like the castle), this woodland has needed urgent renovation. Picking a cluster of pine, I stripped the needles into a pinch between my thumb and forefinger, lifting them to my nose. A habit, summoning memories of small me thrusting moss under my parents' noses insisting: 'The Earth! It smells of the actual *Earth*!'

For years, conifer giants dominated the 825-acre (334-hectare) Fingle Woods. The demand for fast-growing mining timber during the Second World War surpassed the demand for deciduous. What followed was a period of cyclical planting and coppicing of a commercial Douglas fir plantation through to the 1980s and the increasing retreat of ancient, broad-leaved woodland. Less than 3 per cent of woodland in the British Isles is classed as 'ancient'. The Woodland Trust designates this as 'that which has persisted since 1600 in England and Wales, and 1750 in Scotland ... relatively undisturbed by human development ... the most complex terrestrial habitat in the UK'; and my favourite, simply 'irreplaceable'.

An ancient woodland – with all its secrets – takes centuries to establish but minutes to destroy. And if I've understood correctly, once it's gone, it's gone. *Finito.* As I write, the demolition of ancient woodland to make way for England's high-speed rail line (HS2) permeates weekly news bulletins. As though unfairly reprimanded, nature is paying an infinite price to reduce passenger journey time in a project that will require nearly £200 billion of taxpayer income. Rivers are to be canalised. Vital corridors for wildlife are likely to be severed. And 60,000 human bodies were exhumed from cemeteries that were getting in the way. Acres of woodland shaved as though in preparation for prison. The majestic Hunningham Oak stood for three centuries in Leamington Spa until one Thursday (in 2020) when contractors felled it for a service access road. Oh well, at least the speediness of HS2 will give us all more time to grab another skinny latte with caramel drizzle!

Thankfully for Fingle Woods, it's had kinder allies. The Woodland Trust joined forces with the National Trust, and these conservation heavyweights sought to rebalance evergreen with deciduous species, buying Fingle Woods in 2013. So began one of the most extensive woodland restoration projects in the British Isles. The Teign Gorge now hosts one of the largest areas of continuous woodland in south-west England. Attempting to find a space in the car park at Fingle Bridge alone demonstrates its popularity. People flock to bathe in Fingle Woods' leafy spa before queuing for a carvery at the Fingle Inn.

Spiderwebs knitted the dew across the grass at my feet, and I bent the top half of my body at various angles to allow the morning light to escape from behind and reflect off this artistic little array. The usual huddle of 4x4s lined the lane at Mill End – Saturday morning walkers heading into the woods. I had a sudden urge to climb atop a black Range

Rover and take in the morning from its lofty height, but I didn't.

Woods have long provided a rich resource for humans, not least in acting as a giant natural carbon sponge erasing our casual recklessness. Britain's woodlands store 213 million tonnes of carbon, 77 million tonnes of which are held in ancient woodlands like Fingle. Fingle Woods' hunting, gathering, trumpeting past is heralded by the four Iron Age hill forts guarding the valley. Charcoal hearths and stone boundary pollards suggest early woodland management.

A white stag often surveys the gorge on an early autumn morning, his hinds hushed together in the copse below. Honestly, what sounds like fiction couldn't be more real. A condition called 'leucism' will have caused this stag's hair and skin to lose their natural colour. As I continued to stare at the patterns of dew, I half expected Mr Tumnus to appear from behind a tree and escort me with his dusty lamp. Curious walkers can retrace the steps of our medieval ancestors at Hunter's Tor, Fingle Packhorse Bridge, Forester's Track and Hunter's Path – part of the increased footpath network created in the restoration project. There is something enchanting about the whole site, and it wasn't the first time that I lamented my lack of white stallion, noble wolfhound and billowing velvet cloak as I lingered on the edge of the wood. But, alas, modern salmon adventures work better with neoprene, a kayak and a knowledgeable friend.

Joe arrived in a well-travelled silver Volvo. Reliable and sturdy. Inside, it was reassuringly messy. I always trust people who have messy cars. On the whole, they turn out to be decent humans with heads and hearts firmly in place. Joe had two kayaks strapped to the roof of his car. When I saw them, I itched to get going. But first, we shared sadness at the dead fox we had both seen lying on the roadside while we pratted about with buoyancy aids and fastened helmets that made us look like Powerballs awaiting a spin in the lottery gyroscope.

Recent localised downpours had plunged the river into a high spate. The water looked *busy*. The previous week, I had brokered a random conversation with a guy in a pub garden; he happened to mention that the salmon had started running again. The chase was on, and I wanted in.

Identifying a small sandy area next to an eddy current, we lugged the kayaks onto the shore, wrestling with white-water spraydecks, paddles and gloves, and chatting about the route ahead. Feeling like a born-again novice, I was grateful for Joe's wealth of experience on the water, given the various rapids, mini waterfalls and tree trunks I knew we would meet on our journey to the salmon weir. Even though we had walked the entire route in preparation for getting in the water, I felt both ease and apprehension – like making a friend at a new school or knowing one other person at a party. We embarked.

Launching in a kayak is rarely graceful. Sitting in the sandy shallows, we must have looked absurd. Sealed within our spraydecks, the only way to move was to hump our upper bodies forward, every thrust edging us closer to the water and eventual buoyancy. We collapsed into laughter. It was around 3 kilometres to the salmon weir, downstream of our starting point. Clear, rust-coloured water took on a new language around every rock, stick and depth change. Barely paddling, we drifted through a leafy tunnel that drew us in like a capillary – nature unruffled. Deadwood piles on the riverbank were gilded with honey fungus. A kingfisher darted towards us like a brainwave, finding a new perch on an overhanging beech branch. The Teign had outdone itself, and I wanted the whole scene to wrap me up into a riparian burrito.

The wording of the Woodland Trust's Management Plan for Fingle Woods gets straight to the point in celebrating the wildlife that calls the Teign Gorge home, referring to it wonderfully as a 'rich oceanic woodland assemblage'. When reading the list of species you can find here, you could also be reading the list of artisanal coffees a bearded barista will craft for you at his organic, hand-sewn cafe. We have otters, dippers, slender bird's-foot trefoil, woodpeckers, redstarts, pied flycatchers, pearl-bordered fritillaries, dingy skippers, a shot of barbastelle bats, a drizzle of sea trout and of course, a dash of Atlantic salmon (with extra foam).

Teign Gorge aside, I think it's fair to say the world enjoys a salmon. Global seafood consumption has doubled in the past 50 years. Some experts say the UK has entered into a 'seafood consumption crisis'. 'Bit dramatic …' I hear you say as we queue for fish and chips. Even so, our taste (for salmon in particular) has increased annually by nearly 15 per cent. And in 2017, we ate about 100,000 tonnes of the stuff (me included, I confess). Global figures match this pace of demand. Salmon tastes good! It looks pretty on my plate! Full of the omegas and all that good stuff.

Because of its appeal, in Scotland alone, the salmon farming industry plans to double in size by 2030. With salmon's role at the top of the food web, this means that an extra 310,000 tonnes of wild fish will be needed, per year, to feed the salmon. Where will they get them from, then? With the global human population growing by more than 80 million per year, our relationship with seafood is here to stay, but it's getting messier.

Over in the US, a terrifyingly industrious salmon farm in Florida has grand plans to supply more than 40 per cent of the USA's annual salmon consumption by 2031. Closer to home, an investigation of some Scottish salmon farms in 2019 discovered that up to 45 per cent of farmed fish die – often from disease outbreaks – and end up in landfills. 'The huge densities of farmed salmon in the nets provide an ideal breeding ground for sea lice, which provides a poisoned chalice for young, wild salmon to try and survive,' said Dr Janina Gray, deputy chief executive and head of science and policy at Salmon and Trout Conservation. Janina is a fierce, impressive woman – a rare breed in this fishy world. After agreeing to meet over Zoom, I had accosted her with a hailstorm of questions. It turns out the salmon carcasses were shipped in containers to disposal sites as far-flung as Denmark. Yeah, we enjoy a salmon. We just don't want to think about how they arrive on our plate.

Salmon are restless. As fidgety as kids on a long car journey. I mean, wouldn't you be? If you are both officially endangered and officially edible? The status of the UK salmon population is a valuable benchmark for international decision-making. The designation of Special Areas of Conservation (SACs) around Europe have been guided by the status of salmon in UK waters. But our waters are in crisis. Salmon have been swimming in their lowest numbers in UK waters since 1952. A survey in 2020 found *every single* English river (nearly 1,500 of them!) failed water quality tests – just 14 per cent were of 'good' ecological standard. Yet as I write this in 2021, a new report from charity Surfers Against Sewage found that incidences of sewage discharge in UK waterways have risen by nearly 88 per cent since then.

But before we dig into that, I want you to *know* Atlantic salmon. As with so many species under our noses, evolution's delicate fingers have spent millennia knitting a life so rich in purpose that the average FTSE 100 achievement pales in comparison to this fish. An ambitious existence has been ascribed to them. Salmon conduct one of nature's most astonishing migrations as part of their life cycle. I don't envy the endurance they have to muster, but I am in awe of their focus – unwavering and absolute. So were the

Celts, as it happens, generations of whom have associated the salmon with wisdom and power.

Luckily for us, global folklore is richly splashed with salmon. Salmon are allied to the valiant protagonist in his quests, strength in battle and insights into the future. Lost rings, mystical hazelnuts and a divine child all play a part in salmons' heroic portrayal. In Norse mythology, Loki, the god of mischief, once transformed himself into a salmon, infamously leaping into a pool to escape the fallout from one of his evil tricks. A wise move, in theory, were it not for his massive brother Thor, who plunged his giant hands into the pool to catch his naughty little brother. The tapering of a salmon's body towards the tail that we see today is rumoured to result from Thor's legendary grip.

Rising in Cranmere Pool on Dartmoor and leaving us at Teignmouth, south Devon, the Teign is, for many people, their native river. So, too, is it for some salmon and sea trout. Of the two species in the Teign, sea trout are more numerous, which is thought to be because they are a more adaptable fish and currently the healthier of the two. As a result, Teign anglers are usually casting for sea trout. Salmon, not so much. Environment Agency data from 2019 reported 39 salmon catches compared with 538 sea trout. My odds of seeing salmon were, therefore, pretty abysmal.

Once a year, salmon or trout return to the Teign searching for sexual fish-tercourse (I'm never saying that again). Fair play to them after a long journey from the sea. *Salmo salar:* the 'leaper'. Our veritable god of the river. With its glistening, rippling muscles, this athlete is adept at overcoming obstacles, fiercely loyal and an invincible navigator. Surely one to bring home to the parents? The river is its lifeline, sparking the birth of generations of salmon. Once a staple player in all countries whose rivers spill into the North Atlantic, the potential for salmon to realise their purpose has rapidly dwindled. As our little lives have become bigger, salmon have unwittingly taken on lead roles in a modern tragedy.

Cold-water lovers, salmon thrive in only the very cleanest rivers, mainly in the north and west of the UK. As one of Devon's 69 rivers, the Teign should feel flattered that salmon continue to return here. To be chosen by salmon is a huge compliment, and one that rivers should readily accept. Where there are salmon, there is life – our aquatic canaries of the coal mine. The river belongs to salmon. Pie and a pint, scone with jam and cream (*jam* first – go on, sue me), freshers and flu. They go together.

However, a salmon's life cycle is bloody ridiculous. A locked-and-loaded biology lesson, if you're up for it? Other anadromous fish (fish that feed at sea, returning to freshwater to breed) have received similar genetic and environmental cues to adapt this strategy – including lampreys and sea trout. Bear in mind that these journeys can be perilous marathons of more than 6,000 miles between river and sea, with reels of pea-sized eggs laid and fertilised somewhere in the middle – all to satisfy their collective urge to continue. I'm already exhausted.

Beginning in their river of choice during the autumn, anywhere between October and November, the salmon's travels can extend into late February, particularly in larger rivers. Against all odds, adult salmon achieve the impossible and return to the same river in which they were born to close the circle and give rise to the next generation. Thought to recognise the taste and scent of their home waters, returning home renews this extraordinary bond, and the loop continues. It was at this last stage that I hoped to meet the salmon returning to the Teign.

Fat and heavy from up to two winters of serious gorging at sea, these returning adults stop feeding altogether once they hit freshwater, resorting to harvesting body fats built up during their Atlantic voyage. Sufficient fat reserves aren't always guaranteed, thanks to climate change, shifting prey distributions and over-fishing. It may yet be weeks until the salmon find a mate and lay eggs. It's a testing time.

A female salmon will seek a lovely stretch of cold water with high oxygen levels. Using her tail, she scoops a nest (called a 'redd') in among loose gravel on the riverbed. It's only natural that size is important for these girls. The larger the lady, the more eggs she can deposit (this is 'spawning'), with females weighing more than 10 kilograms able to pop out a meagre 15,000 orange eggs each. Go, girl. Such fertility is a game-changer in a hostile world. More eggs provide the foundation for Mother Nature's Life Insurance Policy (terms and conditions apply).

Curiously, in some bizarre reference to poultry, female salmon are known as 'hens', and males are known as 'cocks' (if anyone knows why this is, can you let me know?). Once our hen has laid her eggs, she swims off, and a cock, bright red and dashing in his mating attire, swoops in and releases a white fluid called 'milt', which settles among the eggs, fertilising them. 'Some young males are called "precocious", maturing in their second year while still being the size of a dollar bill and having never gone to sea,' said Janina Gray. 'These males sneak around in the female redds, trying to fertilise the hen's eggs alongside the adult males.' 'Twas ever thus.

The perseverance of an Atlantic salmon is something I wish were more celebrated. Friends, it is *outrageous*. Where their Pacific cousins always die after this spawning effort, Atlantics can sometimes repeat the *entire* ordeal all over again, heading back out to feed at sea and returning some years later. As with anything, though, evolution has been picky, affording only the very strongest, fittest fish this privilege. A world away from their post-Atlantic bulk, by the end of this journey, they are thin, weak swimmers and vulnerable to being eaten, so only 5–10 per cent of salmon succeed in a second marathon. (Many of these are female – just saying.)

Early into our journey, Joe and I passed under an old iron bridge into a deeper body of water. This was our moment to see whether we, the elite mammal, could match the strength of a salmon and paddle upstream. We failed miserably, tiring almost instantly. The water suddenly switched from idyllic babbling brook to impenetrable torrent. We oriented ourselves against the deluge, convinced we were hearing chants of 'Come on, *losers*! Let's be havin' you!' We laughed dispiritedly, shouting encouragement to each other, teeth gritted and shoulders burning, as we wrestled with our weakness in this watery turf. We are no match for this fish.

Tucked away from predators in the gravel, fertilised salmon eggs quite literally chill out for up to 50 'degree days' (depending on water temperature), the water acting as their incubator. The yolk sack within each tiny egg is a crucial lifeline for these infants, their only food as they develop in the redd. Once hatched, this is the 'alevin' stage. As they have yet to acquire any swimming gear, the hatchlings are relatively exposed to passing trout. Once the alevin has absorbed its yolk, it becomes a 'fry' – which, although it can finally swim and look for insect food, will remain in the river for up to three years before leaving home for the sea.

After the fry stage, the young salmon become 'parr' – named for their camouflaging bands of 'parr' marks across their bodies. Spending a few more years in freshwater, parr wait for the ocean's call. Here comes the magic: the salmon hit puberty and initiate their adolescent rebellion against their maternal river, with a spectacular silvery costume change to blend in with their salty new world. This is known as 'smoltification' (excellent word). It's thought the sudden desire for saltwater is triggered by longer days and rising temperatures as spring approaches, which induces a shift in hormones, stimulating bodily transition. Teenage salmon (or 'smolts') cannot adapt to saltwater immediately. As they swim downstream, they loiter near the mouth of the river for a few days to acclimatise, gradually moving to saltier water each day to allow an astonishing series of physical changes to take place. For the Teign, the seaside town of Teignmouth in balmy South Devon is the local bus stop, temporarily housing our restless youths before they make the leap to Atlantic freedom.

'It really is an amazing transformation,' Janina remarked, 'with so many changes happening, they are very vulnerable and sensitive – just like teenagers!'

Having returned to a less exhausting downstream course, something told me things were going too smoothly. There was play in the way the river spoke as I rocked carelessly from side to side, spinning backwards down mini waterfalls, feigning skill. We were moving fast. The current shuttled us downstream, like Poohsticks. Joe stayed up front, testing the route, determining which areas were safe and which I should avoid with various hand signals. I couldn't rely on him reaching me, so I had to stay on the ball. Which, of course, I didn't. The thing about being on a wild river is that events can unfurl very quickly. If you choose to be on, in or under the water, you must accept your subordinacy. Water is untameable.

Joe paused in an eddy, paddling backwards to maintain his position. I sidled up, halting my course by grabbing a branch overhead with my right arm – rookie error. The current snatched the left side of my kayak, pulling me over in a deft flourish. My left hand grappled for a root under the water, frantic for land. My right arm tried to seize the branch above, but the moss that coated it made it slippery. Total capsize – being trapped under the flow of turbulent water, lacking the skill to right myself or escape the tight seal of the spraydeck – was fast approaching.

Joe, realising what had happened, was fighting the river's attempt to plunge him downstream away from me, struggling himself to round a massive rock in an exceptionally messy piece of water. '*I CAN'T HOLD ON, JOE*!' I shouted, my voice tiny amid the rush. Fighting to sound calm and blasé, my shoulders and right bicep screamed to let go of the branch, my core on fire keeping me from going under the water. It felt as though the entire River Teign was now pouring down the back of my neck, freezing, desperate to drag me below. I tried to ignore the wall of my kayak rising on the right, making the whole boat perpendicular to the water, ready to be engulfed. I looked to the tree canopy above, pausing in that brief gap of time that dire situations can conjure, and breathed. Leaves whispered in the breeze. The drama welcomed dark thoughts, and for a minute, I thought that this was it. *Not the worst way to go*, I thought. Rather heroic, even. Certainly very on-brand. But I would rather see the salmon first, please, if possible.

Before I knew it, Joe's arm was in my face, and he wrenched my kayak back over, snapping me out of my head. We stared at each other, breathing heavily. Adrenaline, shock and relief swam over our faces like dappled sunlight. We didn't say anything for a moment, then simultaneously erupted:

Joe: 'Oh. My. *GOD* – are you OK?!'

Me: 'What about *you,* though, are *you* OK?'

Joe: 'You were *so* calm and brave!'

Me: 'No, *you* were so calm and brave! *I hate myself!*'

Joe: 'What?! No, *honestly*, it was *totally* my fault, mate … sorry.'

Me: 'Oh my God, don't you *dare* apologise!'

Joe: 'Sorry.'

Me: 'You *sure* you're OK, though?!'

Joe: 'Yeah.'

Both: 'Need a drink.'

Our drama was a perfect example of the river reasserting authority. We were both shaken. Pale, Joe kept glancing at me nervously as we returned to our course, as though I was about to spontaneously combust. I felt chastened, guilty at being his responsibility and for being rash. I then concluded that we had better bloody see some salmon, else I'd be buying the pints for the rest of the year.

A counsellor once hypothesised that my intense resistance to change stems from grief following my parents' divorce when I was 10. The shock of my parents' breaking up triggered a fifteen-year aversion to things being different, and I freely admit that I tumbled into a decade of denial. (From which I'm still slowly surfacing.) Whatever the root cause, I hate all change equally, whether it's a change in routine, weekend plans or the climate. And don't even *think* of rearranging the furniture – just don't. However, as my mum has rightly observed, I can tolerate unpredictable mishaps that arise during adventures – and trips for this book. I can embrace those changes. Thrive on them, even, yet I struggle to translate that acceptance into the more important areas of life.

For example, I find it deeply unsettling that Atlantic salmon must adapt their entire way of life *and body* in preparation to survive in a sea they've never seen. Like us, salmon are expert at regulating water and salt in their body. And the extraordinary thing is they're able to adjust their physiology to be 'at one' with multiple worlds.

For instance, in the ocean, the salmon is surrounded by water almost three times as concentrated with salt and other minerals as the fluids in its body. This means it constantly loses water to its environment, like how when you're spraying perfume, some of the fragrant particles disperse into the surrounding air and don't remain solely on your skin. This is an invisible (slightly tedious) phenomenon that we call a 'concentration gradient'. Salmon must drink several litres of water each day at sea to offset this dehydration – an endearing paradox. (They don't need to drink anything at all in the river, bar the occasional accidental gulp.) At the same time, the kidneys drop their urine production dramatically. The gills also have a starring role in sorting out what goes where, balancing salt levels via a series of pumps that shove excess salt out of the blood and into the surrounding seawater. The same pumps then let sodium in – which reminds the salmon to *Keep drinking! #stayhydrated*. Back in the river, the reverse is true. If not careful, the salmon will suffer from a lack of salt in this more watery world. So, the kidneys step in again with lots of dilute urine to offset all the water that is now re-entering the salmon's body. This is all very clever.

The time a salmon spends at sea can last between one to two years. Any longer is increasingly rare. They do a lot of swimming around, feeding and (ideally) getting fat. Serious bulking is high on the agenda. It's all a bit lads-lads-lads-getting-massive-at-the-gym for a bit. Then it's time to return home, like good boys and girls.

We can finally answer the question: how the actual hell can a salmon – a fish – navigate an ocean without Google Maps to return to the river where it was born? I'm obsessed with how salmon navigation implements a skill set humans will never attain. As with many animals undertaking migrations – turtles, geese, butterflies – the Earth's magnetic field is a salmon's GPS. Research has found that this instinct and salmon's connection to planetary forces are so strong that this ability remains even within landlocked fish on salmon farms. The theory is that if the salmon were to escape, they could find their way out to sea and back again. When at sea, they course-correct, shifting to align with different magnetic fields. It seems they are just born this way, as even our tiny salmon fry respond to the Earth's magnetic field as they emerge from the gravel. Such is the life of river royalty – humble, vulnerable beginnings in the armpit of a river, with the seed of Hercules firmly rooted from the get-go.

Before my trip to the Teign, I chatted with Dr Jamie Stevens, who leads a research team at the University of Exeter that knows an awful lot about salmonids.[28] The life cycle I've just described follows the first-edition stage directions in the play written by Mother Earth in a rosy world where nothing fluctuates and everybody ~~hates~~ loves Marmite. However, within five minutes of chatting with Jamie, I realised that our salmon are thespians in a much darker drama. A study by the Zoological Society of London (2020) looking at global populations of all migratory fish[29] found a 76 per cent reduction in numbers between 1970–2016, with the most noticeable impact seen across Europe, undeniably linked with humans. You may not be surprised to learn that for every 100 salmon that leave Scottish rivers, fewer than five will return – a right *Macbeth* of a situation.

[28] More jargon. This time, to describe both salmon and trout.

[29] Including trout, salmon, eel, shad.

Listening to Jamie, I reverted to earnest-but-ignorant zoology student – lots of nodding. 'Salmon aren't quite as well equipped as trout, but they are a fish that should, as a species, have the potential to do OK,' Jamie told me. 'They've got a very plastic genetic makeup so that they can adapt to different conditions.' Sounds handy in a world that has warmed by around 0.8°C since the 1870s and (in some places) is predicted to simmer by just over 3°C more into the 2060s. We have been here before, but 3–5 million years ago when carbon dioxide concentrations weren't as high as they are today. Well before 7 billion humans hit the scene and gave emails a carbon footprint. But 4°C of warming in less than 200 years? Oh, dear.

The unique ability of anadromous fish to switch between freshwater and saltwater via a few bodily changes enables them to tolerate temperatures ranging from -0.5–24°C. Such flexibility has helped them to recolonise much of Great Britain, Ireland and further north as the ice receded 10,000 years ago following the last Ice Age. Salmon are good at exploiting new places; they're resilient and opportunistic, making the most of what's around them at any given time. We should celebrate such resourcefulness, yet for some perverted reason, we have actively made life harder for the salmon. Despite exceptional conservation efforts, globally, Atlantic salmon are generally screwed.

'It's the sheer range of salmon migration that complicates things,' said Jamie, speaking casually of the historic feeding grounds off the coast of Ireland, south-east Iceland, sometimes Greenland, as though getting to them were a mere jaunt to Morrisons (or indeed, Iceland). 'Our local populations of salmon are naturally southerly, so they've already got a long way to go. Being at sea is just very risky.' Risky indeed, as recent work by the Environment Agency found that the first few weeks at sea for a young smolt are the most important, with the highest natural death rates occurring during this period.

But even before they get there, the intertidal zone is dodgy. 'Which habitats they use, the routes they take to and from the coast … we know so little about what happens to them at this stage,' said Janina Gray. So-called 'chemical burdens' picked up en route to the sea 'could be fatal'. Sitting back in my chair, I had an epiphany. Another year, same me. Boring people, boring lives. Suddenly, enrolling for work experience with an Atlantic salmon seems like a promising career move. I wanted in on all their secrets.

The fact is that for salmon, temperature and timing are everything, much like the requisite dip on an August bank holiday as soon as the sun decides to behave. Our river and coastal waters have been steadily warming during the past 40 years, with southerly areas including Devon and Cornwall being hit pretty hotly. Sea-surface temperature is more than half a degree warmer in just a few decades. What's worrying is that scientists have measured these temperature changes in both our rivers and the ocean (not just the Atlantic) – and discovered they are synchronised. Double the trouble for salmon – and a rather (bad breath) announcement of 'ALL CHANGE!' on a dizzying scale. I fear global warming channels a similar nerve, ready to declare mutiny against our captain of the river. We should be afraid.

A warmer planet has more energy swimming about our atmosphere and oceans, a lot of which is absorbed by the ocean, fuelling hostile waters and increasingly angry weather. These elements are a more obvious threat than the drift-net fishing that lay a death trap across salmon migration corridors around the west coast of Ireland until it was banned in 2006. As is becoming the idiom, the things we cannot see are the most unsettling. A 2019 study by the National Oceanic and Atmospheric Administration (NOAA) studied populations of Pacific salmon between the Mexican and Canadian border. Brutally, it showed how extreme rains and flooding could flush salmon eggs, alevin and fry from their redds. *Whooooosh!* – a labour of love and survival – gone.

The ocean hurdles of salmon sound beyond taxing, but it's the new obstacles they're facing on their return journey that should worry us because as well as making the weather more agitated, warmer temperatures are confusing the salmon's navigational accuracy. Our master of the compass is a soon-to-be refugee in the wrong neighbourhood. I mentioned that some salmon are fit enough to make repeat journeys, but only 5 per cent are achieving this now, compared with 25 per cent 20 years ago. There's a genuine risk of salmon not being able to return home. Imagine running halfway around the world while dodging murderous bandits, being battered by multiple named storms, eating poorly and getting tangled in the odd plastic bag or stray net. You're finally about to turn into your road, your *glorious* road, to settle down and start a family only to realise your grave error – you are hundreds of miles from home, starving, beyond exhausted, confused, homesick. Alone.

As Joe and I continued downstream, it felt as though we were on a proper expedition. Ever since the earlier drama, Joe was even more alert, talking me through fast-flows and mini-rapids and even getting out of his kayak at the bottom of a C-bend chute, ready to right me in case I capsized again. Strangely, I felt more confident, as though my brush with the water had bonded me to it, and I had passed some sort of test. Into the jungle we went.

As we increasingly realise with many animals, salmon can feel stress. I'm not surprised. Not the short-lived panic of an Aldi checkout, I'm talking about a far deeper kind of stress that can translate across generations. Changes in their surroundings can induce that unwelcome belly simmer (can relate) that can disorientate adult salmon when they're trying to find the way back to their river. 'It's very difficult to link the two,' admitted Jamie Stevens, talking of warmer temperatures increasing these navigational 'errors'. 'But there's no question that the homing instinct can be greatly affected.' I asked Jamie whether salmon could just settle for a different river. But he explained that 'If they end up in the wrong place, it's most likely a mistake.'

This increased 'straying' of adult salmon into the wrong rivers can alter the population structure. I can think of many a Devon village whose community is temporarily diluted by crowds of summer tourists. Naturally, newcomers don't arrive adapted to the village nuances. A formerly Apple-Paying urbanite might become unhinged by the village's cash-only economy. It's a similar experience for the salmon. While not always a bad thing, provided they continue to spawn, salmon straying into non-natal rivers are less likely to have the genetic 'equipment' to deal with that local area. However, stretching off-course can introduce vital genetic variation and is one reason why sea trout may be more numerous. But there is evidence to suggest stray fish can also be less resilient.

Jamie's team from the University of Exeter were involved in a study that, for the first time, found Atlantic salmon from a given river can be characterised genetically. Take the chalk rivers of Dorset and Hampshire – a stunning, rare habitat. England has 85 per cent of global chalk streams. Salmon here are happy and settled, married to that river via a unique 'genetic signature' – a fingerprint if you will. When they return as adults, it's like they are renewing their wedding vows. It's exclusive and precious, albeit a double-edged sword in a warming world where their fidelity is threatened and divorce papers are waiting to be signed.

In Devon and Cornwall, salmon are already reaching a temperature barrier. More often than not, the winter waters aren't as cold as they used to be – they need to be to enable the eggs to survive. For all those people for whom the river is a lifeline, this issue of temperature threatening salmon egg survival can become personal.

Denise Ashton works in communications for the Wild Trout Trust, and we met through Beaver Trust to try and bring migratory fish and rodents together on rivers once more. A keen angler, Denise's knowledge and passion for the freshwater world are gentle but persuasive, even over a crackly landline. 'I've had kingfishers landing on my rod because I'm *in* the water, standing dead still. Herons landing next to me, otters swimming behind me. Those are all the obvious things,' she said. 'But we must understand what goes on beneath the surface.'

For Denise, warming river temperatures are a significant concern. I felt uneasy when she referred to the effect on UK salmon populations as 'incidents', as though salmon were being dragged into some criminal investigation. 'We've already had one incident on my local river in the Usk. We had unnaturally warm temperatures around December 2015 and January 2016, and summer spawning in 2016 was seriously affected.' Denise explained that the maximum summer water temperatures a trout or salmon will survive at (usually 20–24°C) are now often exceeded. But it is increasing winter temperatures that are going to do more harm. On the Usk, the lack of redds and spring fry in 2016 led fisheries scientists to conclude that the water temperatures exceeded the critical upper limit of about 10°C that encourage the salmon to spawn and allow sensitive embryos to survive. 'It's an immediate and obvious way that climate change is having a noticeable effect on salmon,' Denise said plainly.

The temperature of the headwaters feeding a river is critical. In Scotland, Norway and even the southernmost parts of Spain, the rivers are fed by meltwater flowing off the mountains. Nice and chilly, ideal for eggs. In Devon, Dorset and Cornwall, the uplands aren't quite as cold as they need to be. Warming waters from hot summers and mild winters can cause fungus to choke and kill any eggs. Similar trends are being seen by the United States, where more rapidly melting snow in Idaho makes salmon survival more challenging.

'Remaining populations of salmon are very vulnerable because they are often low in numbers,' explained Jamie. 'If something happens, then they've got nowhere to go. It's not to say that salmon might not be able to cope because they probably will if they can navigate further north.' Over the next few years, we may have to lose the more southerly populations of salmon in England, save for a few of those remaining populations in northern Spain, where the Picos de Europa mountains feed rivers.

That may not be so bad, though, if we can 'Make Ready the North!' (cue raunchy battledress and a makeshift horn). Global warming will likely make southern areas inhospitable to salmon, which – although 'a tragedy', as Jamie confessed – isn't an utter disaster. Straying salmon may turn out to be a blessing in disguise because, remember, overcoming hardship is hardwired into a salmon's DNA. If we know that salmon are likely to seek out colder, more northerly rivers over the coming years, then we need to rally the troops and make those places as hospitable as we can. Here in the UK, we're talking North Wales, Cumbria, Scotland, where rivers are sufficiently chilly; rivers that would give you pointy nips even after the briefest of dips. Although Scotland's north-west salmon populations have largely collapsed (thanks to salmon farms) – the south of England has other issues. 'People pressure,' Jamie sighed, 'it's just a big thing. There are many interruptions to their life cycle down here when it's already so complex, so energy-expensive. If a lot of weirs are in the way, it's difficult for the fish to pass, then there may be a slurry discharge to contend with – it's relentless.'

In England, where 2019 data counted 432 people per square kilometre (Scotland at just 70) – the density of human activity along rivers alone is staggering. 'Barriers' to a salmon's migration can include anything from culverts to weirs to beaver dams – they are simply an obstacle that prevents access. We clumsily celebrate the fortitude of salmon when really evolution didn't intend for their life to be so difficult. 'A fish jumping at a weir is not a good thing,' admitted Denise. 'It's a problem. In an ideal world, they wouldn't have to jump up anything apart from the odd natural barrier such as a beaver dam or fallen tree. The irony is these once indigenous river species are now classed as outsiders, invading a changed ecology.

'My heart sinks because, yes, it is perfectly natural for salmon to be jumping over barriers,' Denise exhaled heavily, 'but now, post-industrial, post-agrarian revolution, we have nearly forty thousand barriers in England alone.' That's a barrier every 1.5 kilometres on any given river.

A salmon's tireless attempts to overcome these obstacles are unbelievably demoralising. It is no surprise that we find entertainment in the theatre of nature, it being our little narrative after all. Type 'salmon leaping' into Google, and you'll get dramatic headlines such as, 'Leap of faith as salmon draw crowds!' and 'Salmon put on a spectacular show in Shropshire!' But they don't always succeed. Denise referred to the tendency for scientific reports on salmonid migration to favour publishing data on those that 'made it' up the weir while neglecting to compare those numbers with how many had failed. A photographer on the River Severn once spent 12 hours capturing the same salmon attempting to leap a weir.

The barriers Denise and Jamie had spoken about can also increase the chances of predators eating salmon en route to the sea. Denise remarked on the ease with which we focus on upstream migration. She emphasised how simple it is for everyone to visualise the concept of a big fish leaping over an obstacle *on its way* to lay eggs. 'But the thing we're increasingly concerned about that is often forgotten,' she said, 'is the downstream migration.' Interesting.

Before they embark on their journey to the sea, smolts gather in shoals, carried backwards on the downstream current. 'They're very wary of going over things, so they tend to hold together in a weir pool waiting for the water levels to change, or they just seem to need to gather together and gain confidence to descend in a group.' A clique of young salmon seeking camaraderie on their maiden voyage and overcoming physical obstacles (usually overnight) paints a captivating image.

The weir pool at which Joe and I had arrived is a perfect example of a 'holding' pool. Large and clear, with plenty of dappled shade, weir pools are vital in maintaining stable and cool water temperatures. Pools like these are the calm before the torrent that follows.

But a hungry goosander (a large, diving duck) knows full well that such still waters will likely harbour shoals of smolts, 'and that's perfectly natural and should not be an issue', remarked Denise. Researchers on the River Tweed, flowing between the North of England and the Scottish borders, attached acoustic tags to salmon smolts to track their movements. Data revealed that 80 per cent of them were lost, suspected eaten. Relentless interruptions prove deadly for generations of salmonids – a barrage of swift punctures on their riparian ride.

The irony is that most research and government funding into Atlantic salmon used to be concentrated in Scotland, but it's the fish in southern England that are feeling the squeeze. Further research by Jamie's team in 2011 found that 'people pressure' – in the form of agriculture, pollution, river recreation – can add to the problem of fragmenting salmon populations into smaller, less viable groups. Warmer rivers and seas then step in to exacerbate this problem. Jamie described how often an English river's headwaters lie within prime agricultural sites: pesticides, fine silts and excess nutrients are all part of the package. 'Death by a million cuts,' Janina Gray lamented, 'adding up to a disaster.' She painted a grim reality of excess sediments, chemicals and nutrients choking spawning grounds, reducing not only insect availability (salmon food), but lowering the oxygen in the water. 'We have no idea about the impact of chronic, low-level exposure to chemicals – it's the elephant in the room in current water policy,' she continued. With 100 per cent of English rivers failing, we need to act fast.

Held in the embrace of Dartmoor National Park, rivers like the Teign are less of a concern. The Missing Salmon Project proposes a holistic view, considering all the factors affecting a salmon's journey from river to sea, from climate change to barriers – like a massive nationwide search campaign following an abduction. According to Denise, Jamie and Janina, 'holistic' is the only approach we should be considering now. Sound familiar?

'There's a big collective scratching of heads going on,' Denise mused. 'We assume the hardest part is over when a salmon reaches its headwaters, but we are grappling to understand why we are losing so many of them.' Fantastic, a murder mystery with no conclusion.

As I write, the concept of 'river buffers' is the new conservation buzzword flaunted in meetings like the next company we should all invest in. A 10–30-metre wooded zone between a river and the surrounding land presents a chance to stop some of our problems from entering the river. For some, it's an end-of-pipe solution. But there's no doubt it would buy more time. 'A buffer will reduce diffuse pollution, increase shade, soak carbon, boost habitat … if it's done properly, it will mitigate many of the existing problems,' Denise told me. It offers hope that the entire original cast of A British River™ will return to the stage one day. Not a ridiculous thought when research from Colorado State University by Professor Ellen Wohl decisively proved how an 'active' wetland and river ecosystem that houses keystone species (like beavers) can store up to 35 times more carbon per hectare versus grassland. It's a simple case of rivers needing more space and eliminating elemental extremes. I believe it.

We need to be bold and disentangle all the elements at play to fight for species like salmon. Climate change is not the sole cause. There is no single factor to blame for the increased hardship Atlantic salmon face. Instead, it's more like an entire symphony orchestra of deft, skilful musicians, and climate change has stepped in as the new conductor.

A few false starts. Light hitting the white water flashed silver, spurring sharp intakes of breath and defeated slaps of the water surface with my palm. You must know that salmon are *large*. Some grow up to 1.5 metres long and weigh in at a whopping 40 kilograms. I prayed I would see one of these giants. Joe stayed by the kayaks, watching from a different angle, both of us starting excitedly when acorns dropped from the canopy above, plopping into the pool, the ripples ever so fishy. I was so excited that I thought I might die upon witnessing the phenomenal feat of salmon leaping, that it might just swallow me up and consume all evidence.

I had lost sight of a dipper that flew downstream, and I became engrossed in sticking my GoPro into the seething beery water below in an attempt to capture what I couldn't see, bracing myself against the line of rocks that separated me from the weir. Then, it came. To start with, just the one. A flash of writhing metal. It leapt up for a second before being swallowed by the white water. 'SALMOOONNNN! Salmon! *Salmon!*' I shouted, jumping up and down in the water and cupping my hand over my mouth. Spinning around to Joe, I yelled to ask if he had seen it too, no doubt disturbing all other wildlife.

Against all odds, it had arrived. Even above the roar of the weir, I was sure I could hear the slap of muscle as it smacked the water, losing itself in the turmoil before summoning the strength it needed to repeat. And then another. And another. All three pools in this weir seemed alive with salmon. Leaping, hurling, persisting, propelled forwards and up by their tapered, forked tail (yo, Thor!) – key features separating salmon from trout. No doubt trout and others were part of this fray, but it was all so frenzied it was impossible to keep up.

By now, a small crowd had formed on the bank, attracted either by our whooping or by the geography of the moment itself. An orderly queue of delighted onlookers waited their turn to sit closest to the weir and take photos. We looked around at each other, grinning, cheering, bonding as strangers in this moment, knowing that some unknown cue would soon cause us to disperse and never gather like this again – these fish undoing the knots between us.

We'd catch glimpses of slender tails and pointed heads at the surface, battling the water while preparing for another leap. They were desperately responding to instinct, forever loyal to the cause. We willed them on, fists clenched and eyes scared to blink – and felt utterly useless, of course. One salmon had almost reached the top. The stillness of the pool where I stood was tantalisingly close, our collective gasps audible every time it got hurled back again over the rocks.

It was brutal. I felt like an overinvested PTA-mum on Sports Day. The desire to escape the sideline and intervene was unbearable (as was the noise I made when the salmon made it into the pool in a final burst of magnificent energy before disappearing into the shadows.) Throughout the whole of that autumn and winter, I never saw them again.

Later, back at home, I flicked back through my videos from the day. Some were better than others. Scrubbing through each frame, watching their bodies thrashing the air, I started to see what it was to be a salmon. When I was 19, that same counsellor suggested I describe how I was feeling. I asked if I could draw it instead, sketching a pebble-bed river on a scrap piece of paper. All the other pebbles represented my friends and family, bouncing along with the current – going with the flow. Easy breezy. I was the awkward rock stuck on the bottom – braced against the flow. But I've since found comfort in realising that that's what salmon do. And how natural it all is. That it's OK. For they are survival itself.

In any movie, salmon would be the hero. An embodiment of all the qualities humans admire. Courage, resilience and persistence in the face of adversity. Our champion. My shoulders ached for days after this trip, yet I had the leverage of a boat and paddle to aid me. I always had the option of escape. I could just choose to get out of the river if it all got too much. I *did* do this. Yet for salmon, it's ride or die – the barriers real, the hurdles terrifying: climate change, more than happening. 'We don't need to manage salmon,' Janina Gray reminded me, 'they are incredibly adaptable if given a chance. We need to manage ourselves and our impacts on the aquatic environment – to allow them the strongest chance to survive.' I saw Earth shift gears that day. And I'm unsure whether I've admired anything more. We can learn from salmon.

I'm not going to pretend that seeing a fish leap over some water suddenly made sense of my little life. But it did feel like an accident gone right. And everything happens for a reason. There is a reason why bumbags are experiencing a glorious resurgence. A reason why one can't seem to apply mascara without resembling a choir boy mid-crescendo. I believe that some salmon leapt in front of me just to let me see them do that. Just to let me pop my head around the door for a second and watch them play the role of a lifetime.

How is it that a fish – one of a group of animals famously mocked for stupidity by society – can achieve a navigational feat with such enormous precision when I – supposedly an intelligent ape – regularly think that south is north and need multiple devices to find my way home? Trying to elevate oneself above salmon is a fool's errand. It'll prompt all sorts of unsolicited opinions, and I really wouldn't bother. Atlantic salmon and their rivers were born to run. And while we let them get on with that, we've got a literal mountain to climb. Shall we?

CHAPTER EIGHT

Mountain Hare

In the following weeks England entered a second, and later third, national lockdown, and I quickly realised how lucky I was to run with the salmon in such carefree abandon. Winter's bite made lockdown harder; the dark frame around each day preyed on everyone's patience like a heavy breather over the phone. We looked for light during those months. And I set myself the goal of making the effort to see the sun rise and set as often as possible – two of few certainties on offer.

Then 2021 arrived awkwardly. The sun inched west across my daily horizon, each day rising a little earlier, a little faster. Spring was walking up the path to the front door. Seasons were blending, and I was almost out of time to see a Highland icon flaunting its winter wardrobe on the Scottish slopes. Snowdrifts across Scotland would begin their melt any minute.

I had gained permission to travel but still felt rather sheepish in this particular pursuit as I ventured north for a second time. Barely anyone knew where I was going. I liked that.

In her 1940s classic *The Living Mountain,* Scottish writer Nan Shepherd told us that we will never get used to the mountains of the Cairngorms – the stars of the UK's largest National Park, nestled in the eastern Highlands of Scotland. Nan's wisdom is hard to ignore, and I re-read that section on the train to London where I would catch my connection, finding myself nodding and making little 'hmmm' noises as though I were blagging a third round of cheese samples at the Christmas market. Written shortly after the war, Shepherd tells readers how writing the book offered her a welcome escapism and an outpouring of her mountainous passion pools across every page. Although these are starkly different times, I can relate to the comfort she sought in writing during what must have also been a chaotic time. Astonishingly, Shepherd's manuscript remained hidden and unpublished until 1977, 30 years after she finished writing it. It is now rightly regarded as one of the most celebrated accounts of a landscape and an author's intimate, personal relationship with it.

It had been one of the warmest days for the time of year, and I had even sunbathed before I left home for the station. Strange, for February. Flooded Somerset wetlands all but submerged the tops of farm gates and, as the train sped along the tracks, a grey heron – our enormous wetland antique – took flight, stirring wading birds like blossom in the breeze. The closing day unlocked a bank of fluorescent cloud, and a short-eared owl left its post and drew pace with the train for a millisecond before veering away to the meadow.

Although the usual buzz was subdued, London was the busiest place I had visited for months – probably since my sweaty hustle on the way to Orkney during the previous summer. I often joke about how I feel a 'bit of a Hobbit' in London, leaving The Shire for the big city, more than likely clad in dungarees and wishing I was barefoot elsewhere.

There is a certain *pace* at which Londoners walk, normalising a sense of urgency, no matter what the appointment. One must walk briskly and always look busy for fear of making eye contact. Night was falling, and my next train – the famous Caledonian Sleeper – was departing at 9.30 p.m. Despite having three hours to kill, I adopted this same, purposeful, London stride between Paddington and Euston, with one rucksack on my front, balanced by one on my back (yes, at times, I'm one of those people). I was saddled like a mule but energised to be going somewhere.

Pigeons flocked around a quiet Euston station. Astride a bench outside, next to Burger King, it took me a moment to take in the enormous billboards masking the construction site for the HS2 terminal set to dominate this area. Photographs of token trees scattered the posters in a wash of green. A 'helpline' was advertised repeatedly across all marketing. Aha! Great stuff. Are you a helpline for those concerned by HS2's threat to 108 ancient woodlands or vexed by the irreversible harm to 693 wildlife sites? What's your refund policy on the damaged items? I'd love to know.

My earlier zest at becoming a temporary city girl was superseded by the sheer thrill of boarding what I imagined to be Britain's Orient Express – or – better still, the closest thing to the Hogwarts Express. Trains have an allure for many, but few things can surpass the delight that arrives upon boarding an overnight train in England's capital, let alone the satisfaction of travelling in such a time-efficient way. Sleeper trains had a nine-year decline in popularity until 2018. Perhaps because air travel between UK and European destinations was becoming cheaper, faster and more accessible. But you'll be pleased to hear that the UK is raising its overnight game. GWR is renovating the Night Riviera to Cornwall, and our Caledonian Sleeper has had a £150 million refurb and enjoyed a near 30 per cent rise in sales to date.

Some debate will always surface around whether sleeper trains are greener, given that they accommodate fewer seats than a standard passenger train. However, with air travel emissions rumoured to triple by 2050, I'm happy for my footprint to be train-shaped.

As I write, it's fascinating to read how the pandemic is affecting future travel choices. Dubbed the 'Covid climate effect', a YouGov study for London North Eastern Railway found more than a quarter of participants aged 18–34 have cited Covid-19 as motivating more environmentally friendly travel, with staycations increasing in popularity.

Looking forward to my overnight journey, I had wondered what it would be like. My recurring, fictitious scenario would begin with a wee dram of single malt, followed by a confident hobnob among fellow, fascinating travellers. (I also seemed to have aged by about 40 years here.) Anyway, back in my cabin-for-one, I would spritz my pillow with a complimentary organic lavender spray before retiring to a cosy bunk, hushed into a dreamless sleep by the rhythmic lullaby of the track. Awakened by a soft wee knock at the door and a steaming mug of coffee, I would soon be surrounded by snowy peaks and endless opportunities in the day ahead.

Regrettably, the pandemic (and too many films) had slightly jaded this imaginary version of events. Pillow spray and the dining car were temporarily suspended for a start. However, I struggled to contain my joy upon seeing the gleaming locomotive (I think we can agree that the sleeper is no longer merely 'a train'?). A string of carriages as green as the Caledonian forest were lined with smartly dressed staff holding clipboards, welcoming passengers aboard. I had treated myself to a cabin. Honestly? It was tiny. Rucksacks and boots sprawled the entire floor. But I *loved* it. And I did get a complimentary bar of soap. Pulling away from Euston, I was so excited I may as well have been waved off by Mrs Weasley on Platform 9¾, en route to my new life at Hogwarts School of Witchcraft and Wizardry. This time, a mini bottle of Jack Daniel's served as a tenuous link to my destination. The acclaimed 'Father of National Parks' John Muir once famously said, 'the mountains are calling, and I must go'. Will do, John. And oh, be a babe and throw in some hares while you're at it?

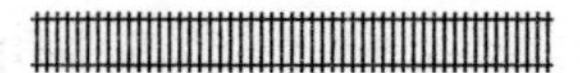

VisitScotland states that 'Mother Nature dealt the Cairngorms a hand full of aces' – an analogy that always lends itself well to nature. Established in 2003, the National Park – named after the Cairngorm mountain range that features so provocatively in Shepherd's *The Living Mountain* – stretches over 4,000 square kilometres of mountain, moorland, river, lochs, glens, waterfalls and forest. This largest region of Arctic landscape in the UK is the unique spread to a rather wild sandwich between Dundee to its south and Aberdeen and Inverness to its north on Scotland's north-east coast. I was reminded of a filming escapade with a couple of uni mates in 2018, where we tensely endured Scotland's infamous battle with midges. Safe to say, this land leaves its mark.

Having never been skiing, the prospect of a 'winter in the Highlands' is one of those things I've been desperate to claim. One of few remaining places in the British Isles with guaranteed snowfall each winter, the plateau of the Cairngorm peaks reigns as the highest, coldest and snowiest in Britain. Five of Scotland's six highest Munros are in this range. And a popular way to use the Caledonian Sleeper is for spontaneous ski weekends for southern city dwellers when the cherished 'bluebird powder' days of fresh snow and light winds grace the forecast.

Just days before I visited, much of the Cairngorms was in a vice of record snowfall. Many claimed it as the 'best winter since 2010 … a *proper* winter, you know?' The average February snowfall per week here is around 12 centimetres, but these weeks saw at least 70 centimetres in parts. It shouldn't come as a surprise to learn that the average number of 'snow days' in the Cairngorms has flip-flopped like this over the past 10 years, with some winters having just two days of snowfall. Since 1997, Scotland has endured its 10 warmest years since records began in 1884. Such variation in weather – and the failure for snow to commit to the slope or frost to postpone the clock – is the new normal. For a country whose climate has leverage over the rest of Great Britain, I sure don't relish the knowledge that my chances of experiencing a 'proper winter' in Scotland one day are likely melting.

Centuries of harsh weather conditions here have ensured that plumages, pelages (a fancy word for fur coat), mating rituals and flowerings are ready to be tested. For only the most resilient and foolhardy, dare I say *hare-brained* (I'll see myself out) can withstand the ferocity of a Cairngorm calendar. Yes, the mountain hare was my weekend aspiration. Naturally, the very name promises an aesthetic that channels a far more Lara Croft-manner of excursion than a sunny cycle to see a butterfly. I adore a sunny bike ride to see a butterfly, but crikey, I was feeling up for a bit of a raid on the February mountainside by this point.

Lepus timidus – apparently, 'the rabbit was nervous', according to whoever named the mountain hare at the time. Come on, judging by the look of your average eighteenth-century scholar (who named many species), wouldn't you be? Not one to succumb to a single label, this hare has garnered several aliases over the years: blue hare, tundra hare, variable hare, white hare, snow hare and alpine hare. This is an animal that has charmed many. The silhouette of a hare is classy. I think we love it. We might be a bit mad for it. At the very least, its long, sprinter's legs and tall, fluffy ears have taught us through the ages that haste and recklessness don't always pay (thanks to Aesop's iconic fifteenth-century fable *The Hare and the Tortoise*).

Please note that hares and rabbits are entirely different species. Unlike the brown hare (and the rabbit), which were introduced to lowlands by some Romans, our mountain hare is a true Scot – as native to the Highlands as the Great Highland Bagpipes. Those among you who have been to the Peak District may well have seen its classic winter white form among the heather, but those populations were brought south from Scotland as game in the 1850s, so they aren't strictly 'native'.

That nineteenth-century translocation was essentially to fuel a fancy new field sport known as 'driven grouse shooting', which was fast becoming *all* the rage, darlings. Swathes of uplands were being volunteered for heather burning and red grouse breeding so that cliques of wealthy, pantalooned chaps could shoot them every August. More on that later, no doubt, but this new land management coincided with releasing a hare or two in the English uplands. However, the core, native mountain hare population resides in the Scottish Highlands.

Classed as a 'boreal' species, the mountain hare is no stranger to the northern hemisphere. Populations hop about the Alps and the Baltic. Although globally it is among the species of 'least concern', its Scottish situation is drearier, with its revised 2020 UK classification now 'unfavourable to inadequate' – 'inadequate' marking the fact that there continues to be a lack of information as to its actual population figures. All this admin has bumped Scottish populations into the 'near threatened' department of the dreaded red list. Whichever way you look at it, it wouldn't take much to prod this species into an irrecoverable spiral.

Fortunately, this mammal has charisma on its side, which (I would hope) gives it a decent leg-up in the campaign to *#BeKindToMountainHaresPlease*. For a start, it's magic. This animal can change colour. Not on demand like an octopus (let's not be ridiculous!), but three times a year, something in a mountain hare's brain says, 'Come on! For everyone's sake – put the other coat on! This one is *sooo* last season?!' An invisible chemical cue renounces one colour for another, ensuring mountain hares stay on track with the latest wardrobe muse.

Colour-shifting aside, mountain hares also have the skill to extract an unbelievable quantity of nutrition from a (presumably) bland diet of poor-quality vegetation. I'm happy for them. Mountain hares actually accumulate body fat over winter despite the pickings being – shall we say – slim. Generally feeding more at night, their menu includes twigs, heather, sedge and grass. Yum. However, where their rabbit and brown hare cousins seem unable to extract the nutrients as efficiently, it's thought the mountain hare has evolved to do so in a bid to get plump and seductive in time for its breeding season, which begins in late January. It turns out that where other animals struggle to survive within the rigour of the British uplands, our mountain hare thrives.

As for that colour change, studies have shown that receptors in the hare's retinas transmit updates on day length to its brain as spring arrives, stimulating a shedding of white hairs and regrowth of brown, which contain melanin. The reverse happens, of course, in the run-up to winter. This process is another of those elegant user instructions from evolution's handbook, where the trio of time, genetics and environmental pressures carefully customise a species to fuse with the seasons, helping it find its seat at the habitat table. Or at least, in an ideal world, they would. Unfortunately, these animals may soon become victims of their own disguise – of their own celebrity, even. Changes that are so natural for the hare and have ensured a relatively covert winter operation out on the snowy hillside for millennia are no longer coming soon or early enough. Biologists are predicting a mismatch of epic proportions to be taking place. Suddenly, winter white is heading out of fashion. (And much faster than it should be.)

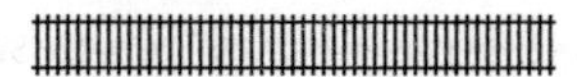

The Cairngorms National Park is one of the last places in the British Isles where it's possible to glimpse a community of nature found barely anywhere else. Species that seem familiar only thanks to folklore, literature and film, or so it would seem. It's easy to ignore the fact that swathes of barren mountain ranges and valleys are not actually the Scottish Highland norm. The Cairngorms show us a different Scotland. Therein lies an altogether different mood. As a landscape, it would be fair to class the Cairngorms' forested, rich ecology as an anomaly.

One-quarter of Britain's threatened animal and plant life can be found here, along with a human population of around 17,000. That's pretty mega. For instance, deep in the ancient Caledonian pine forest, you feel like you are walking around an antique shop, tempted to touch things you know are breakable. Of course, being a National Park, the Cairngorms is somewhat protected from the pressure to conform to the land uses that border it. But we mustn't assume that, just because the park is spectacular, it is pristine wilderness. To be honest, is any wilderness left in the UK any more? But when you're standing atop a Scottish peak woohoo-ing as free as an eagle with the wind whipping your Netherlands, yes, you *feel* wilderness in your literal soul.

However, the fact remains that Scotland, in its near-entirety, has been the victim of a complicated assault of the forester's axe, the herbivorous grind of grazers, the obstinacy of shooting estates, and the engine of leisure. Genuine relics of Scotland's ecological prototype are thus hard to find, and those that remain are introverted and recessive. This has been the reality for thousands of years, as we'll see. For now, though, give yourself permission to indulge in the Cairngorms' moss–heather colour scheme amid a mood board of red squirrels, snow buntings, ptarmigans, capercaillies[30], golden eagles, white-tailed eagles, Scottish wildcats and mountain hares all roaming the Cairngorm plateaus and pines. (Some more than others.)

So much for the lullaby of the track. The fragile hours between 3–5 a.m. kept me in a deranged state of wakefulness. As though conducting some giant model trainset filleting the British night, I dreamt that my toddler nephews were laying out their wooden train tracks for us in real-time, piece by piece, responsible for every jostle and clatter. The approach into Aviemore was spectacular, however, and made my journey to Euston station just a few hours ago feel like it happened in a different year. Nestled in a duvet cocoon and peering out of my little carriage window, my whimsical vision of the sleeper train

[30] Snow bunting: a gorgeous little bird; breeds on bare high mountaintops; very scarce. Ptarmigan: (silent 'p'), a resident highland grouse. Capercaillie: huge woodland grouse; native to Scottish pine wood; at severe risk of extinction.

was finally materialising. My thoughts were silenced by a nomadic scene revealing moorland and heathland with snow-capped peaks beyond.

As I stepped out into Aviemore feeling hungover and tired, it was clear that the entire town was hibernating. Tesco and the town bakery were to become regular supply depots over my weekend. The Cairngorms' rural economy leans heavily on tourism and diverse local business portfolios to make a small, modest but attractive town housing a friendly community bound together by the outdoors.

Refuelled by coffee and an egg bap, eagerness resumed, and I stood in the empty street admiring the entire Cairngorm range beyond, brilliant white in the early sun. I wanted to be right up in those folds and ridges immediately, please. Low and colossal in the sky hung the snow moon. Being the second full moon of the year, its name marks winter's final days. How utterly bonkers it is to think that Neanderthals, Saxons, Romans, plesiosaurs, woolly mammoths and dodos will have all seen this moon too.

Already, I could see snow cover had retreated extensively. Most hillsides had shuffled into a piebald appearance. Birds entered the final rehearsals of their dawn recital. Shielded from the wind, the warming sun relaxed me like an unfurling fern. Spring's campaign was gaining support up here too. For the outdoorsy, the Cairngorms might leave you quite dizzy with possibility. There is an endless amount of fun to be had here: trails to wander, peaks to climb, forests in which to bathe, tree roots over which to stumble. I was lucky to have time to spare that day, the following being the big one: #missionmountainhare.

Dumping my things in the only hotel open at the time, I surrendered to the hotspots. Aviemore has an excellent local bus service, which can whizz you around to most of the recommended sites. A walk above nearby Grantown gave me red squirrels, treecreepers[31] and bullfinches. High above flat forested valleys, I followed the twisted bark of an ancient Scots pine with my hands. Age had hugged it into a stiff lean-to with the prevailing wind, where it now rests in a weathered, settled state of mind. With its reddish, gnarled bark and blue-green evergreen needles, Scots pine is the UK's only native pine. Some have even made their way to the pebblebed heaths back home in Devon. Stretching up to 35 metres in height – with most of the leaves clustered towards the crown – these majestic native beauties can easily live for up to 700 years. Fondly known as 'granny pines', these ancient trees offer sweet sanctuary to fellow animal and plant rarities, including the pine marten, wildcat, crested tit and some orchids.

Beneath the trees, light dappled a loamy carpet of fallen needles, like the quivers of sunlight that reach a reef. From the classic (Instagrammable) perch of Farleitter Crag, Caledonian pine forest dominates, framing lochs that glint in the sun, as though coins in some enormous Earthly wishing well. I time-travelled through the extraordinary Abernethy Forest National Nature Reserve. Dense, ancient understorey hugged the path, its loamy carpet flexing beneath my feet. You see, this is how it was meant to be. This is a place that is meant to be *felt*.

Although now a 'priority habitat' under the UK Biodiversity Action Plan, the Cairngorms' native pinewood is part of the remaining 1 per cent left in Scotland. Briefly mingling with the storyline of our Neolithic farmers 6,000 years ago, those who didn't elope to the Orkney islands stayed and colonised, farming goats, sheep and cattle. The post-glacial retreat had tipped Scotland's tree cover into an all-time high, approaching the climax of rewilding delirium:

[31] Not, as I once thought, a type of stubborn poison ivy. It's a tiny bird with a curved bill, which spirals up trunks, feeding on insects.

willow meadows full of grazing elk flanking rivers that host salmon and bears in a wholesome tussle, not far from beavers convening their wetland alliance. Deeper into the wooded margins and wild boar dodge prowling lynx and wolves; and dwarf birch, willow, hazel, aspen, alder, juniper, scrub and Scots pine fill complex gaps. Life at this time was brief, raw and unconfined.

Later centuries brought climatic shifts of colder, wetter weather. Peat bogs capitalised on areas of tree retreat. Treelines slipped down the hillside. Three thousand years ago, Scotland's canopy was already undergoing deep fissure. Much commentary exists on the human fetish for tampering with trees in Scotland that occurred over the following years. Devastating mass clearances pleased agriculture, world wars, industry, deer stalking and grouse shooting. Faster-growing Sitka spruces outshone the pace of our native pines in this rush for timber.

In a 2015 interview in *UK Hillwalking*, the writer and conservationist George Monbiot conveyed a heavy heart as to the treeless state of the British uplands. He told us these places, 'which would otherwise function as our great wildlife reserves … where hardly anyone lives … have even less wildlife than the places that are intensely habited and farmed'.

I had arranged to meet up with a mate of mine, James Shooter, a widely respected wildlife photographer and guide who I've gotten to know through my work at Beaver Trust. James lives just outside Aviemore and kindly offered to show me a favourite haunt of our mountain hare. Now I'll come clean here – I hired a car for the day. You see, one of my intentions for this book was to create it with my mind consciously (and constantly) focused on low-carbon travel options. My ideal choice – carbon-*neutral* travel – is a whole other feat and frankly unrealistic with my budget and limited time. Travelling 'the low-carbon way' is about doing whatever we can to minimise our carbon footprint. With some wildlife, trying to get a guaranteed sighting within your only full day on their local patch requires a brief dual-carriageway stint followed by a 10-mile single-track road into a remote glen that's well off the bus route. The social distancing guidelines still in place also ensured that carpooling with anyone outside my bubble was a total last resort.

Anyway, just outside the Cairngorms National Park, this glen lies among the spectacular Monadhliath Mountains. Not the only place you can spot mountain hares, but it's home to one of the best-studied populations in Scotland (with the added treat of being part of a traditionally managed deer estate surrounded by grouse moorland).

As we laced boots and zipped up jackets, James half laughed as he confessed, 'Between you and me, much of Scotland is ecologically buggered.' I respected his candour. Increasingly, there's no other way to describe the state of things. The rising 8 a.m. sun was beginning to unmask an impressively huge but sparse glacial valley. It looked worn, weathered and dry. I noticed the slip of the tree line, and how in this particular valley, intense grazing over the years has effectively destroyed any opportunity for new growth. Sparse birch framed the lowest thirds of the hillsides, and James told me how this indicated years' worth of deer-browsing pressure. Lack of tree cover has exposed the river berms[32] to erosion, destabilising banks and throwing the flow off-kilter. 'What we're seeing is the

[32]A term to describe a sandy or gravely ridge parallel to a river's shoreline. Or, if you're a mountain biker, you'll be familiar with whipping around dusty berms (sharp, angled corners) as you shred those trails and #sendit.

bare bones of the landscape,' sighed James. 'Imagine how stunning it would be with more birch...?' The truth is, I couldn't.

But I quickly realised there was also more to it. This particular estate is reducing its numbers of red deer in an admirable bid to rebalance the natural ecology. The restoration of relict scrub, lost tree cover and blanket bog is key to their plan. Leading by example, this estate demonstrates to other landowners that, yes, a landscape's scar tissue may be brittle, but it can still provide a firm foundation for new growth.

We walked along a flat track towards the mountains, bedded with the kind of coarse grassland that has weathered many a gale. Feeling dizzy with joy for being back in Scotland, I imagined hares already clocking our movements. The sun grew in strength. It looked like the snow was melting before our eyes. A random rectangle of pine plantation remained. Rather 'manscaped', as it were. 'So, yeah, we've had the best winter for 10 years,' James said, squinting up at the sun as it caught the glen, 'but it's getting rarer – we should have loads of snow every year, but it's just not happening any more.'

One of the things that *is* happening, however, or that was due to kick off around the time I visited, was the mountain hare breeding season. Baby hares (leverets) – usually one to three per litter – are expected to be up and about from March to July, meaning that February is a crucial time for a bit of *Lepus* tête-à-tête. Thus begins one of the most unmissable mating rituals in the animal kingdom: the hare box. Please welcome to the ring, our first competitor, the male mountain hare – the buck. As tends to be the case, he has realised his sexual peak in advance of his lady – the doe. Gone is his usual reserved, timid demeanour because a tidal surge of testosterone has breached the flood defences. All he can think about is dodging a well-placed hook from his (ideally) equally amorous doe and bagging a quick shag. But first, he must catch her.

Perhaps assuming that she wants to be chased, he reaches speeds of up to 40 miles per hour, so desperate is he in pursuit of a romp. But she is resolute, and I love her for it. This whole flirty catch-me-if-you-can situation can begin to grate on a doe – she's got enough things on her plate as it is! And at the crucial moment, she is likely at the very end of her tether. The prospect of a cheeky interlude is the final straw for what has no doubt been a rather tiresome frolic, often with more than one buck. So she whips around, using her strong forepaws to flail and rebuff the buck's advances. It's an extraordinary courtship because where most species fight for a mate with members of the same sex – most often male to male – mountain hare boxing is almost always between males and females.

A video I once saw conveyed a particularly juicy exchange between a pair that likely belonged to the same population I was hoping to find with James. The unleashing of a rapid-fire of feet and forepaws reminded me of those trivial scraps between siblings. In our hares' case, it's not without cause, however, as the doe is usually communicating one of two things. She's saying, 'I am not ready to mate, kindly off you f@#k.' Or she is assessing the buck's agility and potential fitness as a father, testing him in a gruelling physical challenge. Safe to say, she will let him know if she likes what she sees. I appreciate the sense in it, given that female hares ovulate during sex, so pregnancy is pretty much guaranteed. Casual romps while hunting for 'the one' is not really an option for the doe unless she wants instant baggage. Either way, it's thought that this entire saga gave rise to the phrase 'mad as a March hare', and I can see the value in bringing a little more of the doe's well-meaning fury into modern dating culture.

Before I left home for the Cairngorms, I caught up with Dr Scott Newey, an animal ecologist from the acclaimed James Hutton Research Institute based in Scotland. Scott has spent the best part of the past 20 years working with mountain hares and a range of other upland species. He is no stranger to the mountain himself. His path started in mountaineering and outdoor education before meandering into animal population studies. As we spoke on Zoom, I pondered aloud to him whether the mountain hare was the epitome of his two passions. He laughed. 'They are an amazing species, but on the mountain? They're always happy when I'm at my wits' end. I never personify the hares – they are absolutely their own thing.'

Teaming up with leading international researchers, Scott co-authored a remarkable paper for The Royal Society in 2020, exploring whether climate change is causing mountain hares to be the wrong colour. As we know, 'phenology' describes seasonal changes in the natural world and, in this case, refers to the colour shift in a mountain hare's coat. Put simply, it just so happens that being white in winter amid an increasingly snowless, warming mountainscape is a bit of a problem.

I mentioned earlier that in Scotland, the mountain hare is now a 'near threatened' species. Scott's research spotlights some unsettling truths. It presents a rare case study of climate change showing itself and offering examples of how it will affect wild animals. Thus far, this 'climate change' character has lingered in the corners, down side streets and alleyways, wearing dark glasses and generally being oh-so-incognito, wielding its power with a mere flick of the wrist and a crack of the neck. It's only just warming up, as it were. But with mountain hares and this new research, we have at last been presented with an unlikely opportunity to glimpse our invisible assailant. You ready?

'Between the 1950s and 2016, we estimated there are 35 days less snow cover in Scotland,' Scott began. (That's around half a day less snow each year for 60 consecutive years.) 'Climate conditions are far less reliable and consistent than they used to be.' The 35 snow days that Scotland has lost are the pinch point for our hares because it means their camouflage doesn't work when they need it to. Instead, it epically fails them, creating a 'mismatch' that exposes white hares on a dark mountainside in a very real and perverse way. I hate it, yet I am fascinated. A short period of annual mismatch is, in fact, standard and to be expected. Each spring, the snow melts while the hare is still in a state of transition to its browner coat. But Scott's paper presents the fear that mountain hares are now exposed for longer, beyond this seasonal norm. Their appearance on the melting hillsides is as blatant as a clumsy lie.

One thing also worth noting is that mountain hare populations in Scotland are incredibly valuable to science. As we know, not every species has the luxury of being subject to long-term scientific observation. I'm pretty sure every UK dung beetle would envy the dataset of mountain hares. Historical observations of wild mountain hares gathered between the 1950s and 1960s are thought to represent the most detailed survey of population fluctuations and changes in moult (coat) of any animal. Using these data, Scott and the team repeated the field studies conducted during this time to try and see whether the hares are adapting to warmer winters with less snow by changing the timing of their annual moult. They wanted to determine whether the hares are getting ahead of the climate game.

Dr Marketa Zimova has been studying snowshoe hares in the United States since 2009. Populations in Montana, especially, have been well-observed for their camouflage mismatch. This species, *Lepus americanus*, is more petite, but just as white as our Scottish mountain hares. An evolutionary ecologist at the University of Michigan's Institute for Global Change Biology, Marketa investigates animal adaptation and rapid evolution in the face of climate change and habitat loss, and she led Scott's study with Scottish mountain hares. She is another one of those young female scientists who just seems to smash the stereotype. I wish I had met her in person.

Over the phone, I asked Marketa what she thought of the results. 'We were really surprised,' she admitted, 'because they just haven't adapted! We were hoping the hares would have adjusted the timing of their colour moults to match the decreasing duration of snow cover, but they haven't changed at all.' This discovery does seem odd, given one of the main concerns is that white hares against a browning backdrop makes them (I imagine) about a billion times easier for golden eagles and other predators to spot – and catch. 'Basically, we have seen strong effects of camouflage mismatch in snowshoe hares. A lot more of them die; they're just so obvious, and the chance of being eaten increases quickly, by more than seven per cent when they are colour-mismatched. It's crazy.'

Yet this consequence of camouflage mismatch is pure logic and is seen in similar mammals around the world. During our conversation, Marketa mentioned that she and other researchers are seeing the same trends in other species, such as Arctic foxes in Sweden and weasels in Poland. Such research presents more evidence of entirely different animals operating in completely separate systems subject to precisely the same – gigantic – climatic stressors. A growing chorus is chanting, 'Maybe you saw me. Now you definitely see me' across the northern hemisphere. The climate is changing too quickly – evolution is responding too slowly. It's almost as though we need evolution itself to evolve. I suddenly feel very small.

Back in the glen, the silence that cloaked the valley was rare, and I knew it, lapping it up like some parched dog. Many rocks dotted the hillsides, most of them covered in lichen – a dead giveaway to brilliant air quality. Mountain hares can often look very much like these rocks. James told me how the off-white ones 'are the buggers to spot'. To conserve energy, the hares will often huddle in a scrape of ground or ex-rabbit burrow, sitting still in a form and on the watch for golden eagles and other predators. 'Yeah, so I've definitely stalked boulders in the past,' James laughed as I excitedly pointed to a rock.

But then, two ears appeared behind a small tussock, sticking straight up like two glorious furry fingers at the world. Greetings, hare! Very well played. White and grey, greyer than I expected, but the spring moult can time with the hare box. I had finally remembered to borrow my dad's ancient binoculars and put them to better use – raising them to see our rascal.

Impossibly furry, the grey guard hairs on the outer part of the main coat *did* offer a rather bluey tinge. Its almond-shaped eyes were half-closed, either shielded against the breeze or set in some steely attitude against eagles. Or, as I preferred to think, it was just waking up and feeling grumpy. It was bigger than I had expected, too, its long legs crimped underneath its plush coat. The hairs of a hare's main pelage are hollow, trapping air like the fibres in our synthetic insulation jackets. Below this, fine and downy fur is tightly knitted like woven silk, offering enviable warmth. There was no denying I had an immediate crush.

One person who has arguably spent more time with the hares around the Cairngorms than anyone else in Britain is wildlife photographer Andy Howard, author of *The Secret Life of the Mountain Hare*. Along with a few other professional photographers ('Team Hare'), Andy has uncovered some astonishing insights into these animals simply by being patient. Predicting the yawns, stretches and toilet breaks are all part of his skillset (for hares, that is).

Andy's knowledge has been influential to scientists like Scott and Marketa. For a start, Andy's incomparable photographic library of the hares over the years challenged common thinking that mountain hares change their pelage twice a year. Intrigued, I called him when I returned from Scotland. 'Science rarely collects photographs,' he commented, 'but through being little old me with my camera, I support evidence showing they moult three times a year.' Despite debate continuing around whether a third moult had officially 'stopped', the potential for mountain hares to have winter, spring/summer and autumnal coats is now accepted scientific knowledge. A neat little capsule wardrobe. Lovely.

I was curious to hear what 10 years of sitting in freezing conditions behind the lens had done for Andy, save from yielding some of the most stunning portraits of these animals taken to date. 'Well, they're the most boring animals in the world and also the most entertaining,' he continued. 'It's like going to a party, and you go into a room of strangers, and you don't know anybody. But then, over a period in the evening, you get to know the ones you like, the ones with the best characters, the jokers, the ones you want to spend time with. It's the same with hares. Oh, I also talk to them, of course. For hours.' As someone who has grown up in relatively constant dialogue with myself, my stuffed animals or the real pet animals that have come and gone over the years, I smiled at this rather outlandish declaration of kinship with a fellow mammal. 'Being with the hares is spiritual … there's no doubt about it.' You know what? I get it. After all, data, regressions and hypotheses aside, isn't that what it's all about?

We hiked to the snowline and over the peaks, an offshoot of the river carving a deep gully to our left – one hare down, apparently many to go. I wanted more. Having been lucky enough for mountains to form a regular part of my life, my reunion with mountains was overdue. On an adjoining peak, a trio of red stags caught our scent, their big, beautiful heads reading our intentions. No doubt they had already seen us while we were still faffing around on the valley floor.

It's no secret that mountain hares are a favourite snack for the golden eagle. It just so happened that local birders revere the glen we were exploring as a top spot for raptors. James recalled how an eagle once flew high overhead, and he could see the hare's ears twitching, as though they were its long-distance radar detecting the enemy before visual contact. Golden eagles are somewhat re-staking their protagonist claim in some regions of the Monadhliath Mountains, soaring the skies in their highest numbers for more than a century, according to the RSPB.

A chaotic and strained mix of illegal raptor persecution and driven grouse shoot management has been the leading cause of eagle downfall. Despite the boost being tentative, the stabilisation of any top predator following years of decline is always, *always* good news. Life is dead without prey. 'Numbers are more abundant now,' said James, 'but if we want eagles to stay, ideally, we need more hares in the mix.'

Whether current hare numbers are sufficiently healthy to support other Highland species remains to be seen.

Confusing, I know, to fear for the hare's life amid diminishing snow, yet accept that they are part of a larger chain that also needs feeding. To date, there has been a notable lack of dedicated monitoring of the actual population of Scottish mountain hares. But as I write, a new national survey of mountain hares has just launched. One method of 'night-time counting' aims to observe hares at their most active. The Mammal Society's 'Mammal Mapper' app encourages the public (a surprisingly helpful cohort to enlist!) to record any hares they see while walking in Scotland. Both endeavours will undoubtedly offer invaluable insight into where they are, where they are not and therefore, how worried we ought to be.

Back on the hill, the wind whipped us into shape, and all this eagle banter suddenly yielded not one but two. As luck would have it, here was the bird I had only ever read about. With a territory as large as 200 square kilometres to maintain, a sighting felt as overwhelming as I imagine it would be to stand in the same room as Barack Obama (or any of the Hemsworth brothers). Wingspans of more than 2 metres negate the need for binoculars – with this bird, the naked eye is finally enough. 'Look at the *WINGS* on it ... *gawwd*!!', we shouted, so massive and black it was against the blue sky, soaring over infinite ground, yet barely a wing beat to show for it. Another eagle swooped into view. Suddenly, a tiny arrow of a kestrel darted underneath, plunging all three beautiful beasts into the same frame. I could be anyone, nature lover or total agnostic, and still want this scene tattooed onto my retinas with immediate effect.

Right. Driven grouse shooting. A topic I've avoided that would be unwise to dismiss. Scotland is rare in that while its climate and biodiversity are changing, much of its land-use has been slow to catch up. Digest this: more than half of Scotland is owned by just 500 people, representing possibly the most blatantly unequal land ownership of any modern nation in the world. For nearly 800 years, most of Scotland existed as a 'feudal' land, which essentially means land with a 'fee'. A hierarchical jungle of land tenure was held initially by Scottish kings and powerful tartaned clans. You might be surprised to learn that the Land Reform Act only abolished this ancient system in 2003. A few years and several more acts later, and 'the land thing' in Scotland remains murky. Yes, it's easier now for communities to have a say in who and what goes where, but matching land plots with their rightful owners is still proving a throbbing headache. Tetra Pak founders hobnob alongside the heir of Lego. It appears that everyone and their mums would like a dollop of Scotland.

Since the 1880s, driven grouse shooting has become quite the August hobby among the well-to-do. I mentioned that the number of mountain hares is uncertain in Scotland, but some hares are doing pretty well in areas with grouse shoots. Strange. I probed Scott's brain: 'It's hard to answer ... but the circumstances we have for the hares are really unusual.' He explained that the hares in boreal Scandinavia occur at low densities – say two to six hares per square kilometre. 'But on heather moorland managed for grouse shooting, we can have two hundred or more hares per square kilometre at times,' he said. Well then. Why all the fuss about this camouflage mismatch? Surely we can just pack up and go home if there are plenty of hares to go around?

I wish it were that simple. Landowners and gamekeepers manage around 15 per cent of Scotland for grouse shooting, and large areas within that remain vastly unregulated. More comprehensive licensing schemes from the Scottish government are a way off yet. Two types of shoot exist: 'walked-up' – with four to eight guns and dogs to flush the birds out. Then we have the more intensive 'driven' shoots – bigger, 'better', more boujie. Strangely, our hares can be more associated with these. Regular moorland management such as heather burning and predator control have afforded hares respite from typical (um, *natural*) hindrances like a hungry fox. Save for the odd golden eagle, hares on grouse moors are somewhat spared the harshness of a genuinely wild mountainside. Living on a grouse moor is their own rather stuffy national park. I can't help feeling like it's giving us all a false sense of security.

Before we get into all that, I must mention a big *thing* that's been happening to hares in the past few decades. Cited as some sort of parasitic 'middleman' between grouse and deer, hares host the ticks that spread the 'louping ill' virus to grouse. Despite limited evidence to justify these actions, thousands of hares have been shot and killed. Ticks! BANG! The heather is suffering! BANG! Lovely white fur … also BANG! People killed an estimated 33,500 hares in 2016–17. Let's just acknowledge that this is one of the most complex and heated conservation debates in the British Isles. People have also justified hares 'overgrazing' as reason enough to shoot. 'Of course, herbivores occurring at such density is going to have an effect on the landscape,' Scott mused, 'but our work suggests that the annual biomass production of a healthy moorland far outweighs what a very high population density of hares can remove through their grazing.' Muse indeed.

Soon the public got wind of the fact that the mountain hare was at the mercy of the gun – and they weren't having any of it. During my research for this chapter, I quickly realised that Keeping Up with the Hares may well replace *Keeping Up with the Kardashians*, their legislative merry-go-round being the new series to binge. (Netflix, take note.) It's as much of a social conflict as it is a wildlife one. I'm finding it confusing and am slightly drowning in the nuance. Still, I think the main takeaway is this: finally, a shiny, new 'protected' status for mountain hares now exists throughout Scotland, in a late amendment suddenly introduced to Scottish Parliament by the Scottish Green Party in autumn 2020. As serendipity would have it, this bill came into play the day I returned from the Cairngorms. And, as of June 2021, NatureScot announced that licences could only be granted to applicants for specific purposes, excluding the sport shooting of mountain hares. About bloody time.

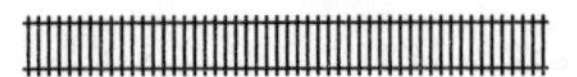

Midday and the sun felt as fearless as we did. Full of vim, we began a new climb up a track used for walked-up grouse shoots. Red grouse frisked among the heather, their hilarious cackles punctuating the cool air. Mating season was creeping up on them too. Grouse butts, in which shooters could conceal themselves, were stationed at regular intervals around the track. Cresting the hill, the white peaked meringue of the Cairngorm range rose in the distance. Snow was more stubborn here, occasionally pulling me, legs flailing, into its soft drifts – a stark contrast to the southern hills which rose above a bare, snowless brown-scape. Years of controlled cyclical heather burning have tessellated this view. Naturally, tiger bread was the first likeness that came to my mind. Breakfast felt a long time ago. 'Peat "hags",' James said grimly. 'That's how we describe the erosion that can be caused by overgrazing or burning, exposing the peat to the elements … often drying it out.' As I looked around, too many hills had this pattern – they looked like a monoculture of tectonic plates shifting against forces too enormous to resist.

Peat is something we do not want to muck about with. Even though it stores more than 3 million tonnes of carbon, we have sacrificed British peatlands for garden compost, forestry, grazing, drainage and this pretentious gamebird pursuit. A window of legal burning occurs annually between October and April as part of moorland management. However, the fact remains that this consistent burning of heather, and subsequent peatland erosion, is increasing by more than 10 per cent a year. Necessary? Hugely debatable. Its future? Uncertain. Release of carbon? Guaranteed. For peat's sake, Scotland's peat alone harbours around 25 times more carbon than all plants across the UK – including trees. I'll leave it there.

Another hare. This time, impossibly white – starched to perfection. 'Daz-white', they call it. Several beats ahead of us, it darted across the path, then hopped behind the grouse butts. Make no mistake; these guys are rapid when they want to be. Before long, several more hares appeared, each as pure as the next. They were like a row of lightbulbs beaming across a dark land, flashing some sort of rudimentary signal we're grappling with decoding. I could see one seated in a furry little ball, shining like the North Star. Four, five, six. The thing is, I couldn't *stop* seeing them. As if they're now shouting, 'Here! I'm over HERE! *Hare* I am! It is I!' Much as I adored watching them, I so wished to be mistaken. Winter shouldn't look like this.

I wrote earlier of the surprise with which Marketa and Scott received the results of their study, so ready were they to find that hares were adapting to diminishing snow. As controversial as driven grouse moor management has been, the fact remains that we may not have as many hares in Scotland without it. Run with me. We know now that controlled burning and predator control are the primary tools in a gamekeeper's toolkit to ensure happy, plump grouse and happy (plump?) clientele. Scott explained, 'the hares seem to actually *benefit* from predator control of corvids, foxes, stoats and weasels, but because of this, we don't really know how important natural predation is to mountain hares in these areas'. He described how hares persist at lower densities in places where natural predator-prey dynamics are not tampered with. Nature's original intention? 'Our studies of hares on grouse moors are thus confounded – it's a real challenge for science.'

Perhaps this all might become clearer if we return to Marketa and our American hares. 'Don't forget that snowshoe hares still have a full predator system in place: communities of lynx, foxes, coyotes, weasels, birds of prey,' Marketa reminded me, 'so as soon as a mismatch of camouflage happens, the chance of being eaten increases quickly.' So, what's happening on grouse moors in Scotland is that we're meddling with their potential to evolve because by removing the majority of their predators, we are releasing a vital selection pressure that would, in theory, trigger an evolutionary answer to the problem of sticking out. Nature made mountain hares the right colour for each season, but our meddling has painted their fur in an unseasonal hue, and now hares are unwittingly wearing the wrong shade. Oh, honey.

'It's the most likely explanation for why they have not evolved to match the 'new' shorter snow seasons,' Marketa admitted. She considered how wonderful it would be if we reintroduced more golden eagles and other key predators that have been lost across much of the Highlands. But this could be bad news for the hares, as predators will find them more often. 'To restore natural systems successfully, we must first understand *all* the players in the game.' Holistic with a capital 'H'.

As we sometimes see demonstrated by a particular demographic in their late 30s, humans enjoy experimenting with artifice. Botox and cosmetic surgery might be a ploy to defy ageing, but grouse moorland predator control feels like a smokescreen disguising a much harsher truth. Selection for an advantageous trait – which in the mountain hare's case would presumably be for an alternative to winter white – is more likely to lead to an evolutionary shift in large, connected populations. We want this to happen.

I asked Marketa about pace. 'Everything used to happen a lot slower – things had time to adapt. Stressors have always existed, but now ...' she sounded burdened, 'there is rapid climate change, habitat loss and fragmentation, overexploitation, invasive species ... it's an onslaught.' Playing with nature this way risks contaminating evolution while denying animals like the mountain hare the chance to fight back and move forward. As the master puppeteer at the top of the food chain, we feed our hunger while fuelling a fire we don't know how to extinguish. A dangerous combination? But the thing is, operating in the short term is just too bloody convenient. 'Nah, I'll be gone by the time things get bad!' we say. But, mate – what if you're not?

Thinking back to some of the hares, especially the sitters, they almost looked like they couldn't care less – very much a 'so what?' sort of vibe. That frustrated me enormously. Like, they're not even trying to sort out the fact that they stick out like an LED bulb! But maybe it's not a problem for them. Evolution is energetically costly and must be worth it for a species to stick around in the long run. Humans would never have bothered to walk upright if it wouldn't have facilitated world domination. Kim Kardashian wouldn't spend an average of £130,000 a month on herself if it didn't help her body break the internet. For Scottish mountain hares, their mismatch in the contrived world of a managed moorland might not threaten their survival. We know that their vulnerability to predation may increase. Unfortunately, plans to restructure Scotland's relict predator community (lynx and wolves) encounter repeated hurdles. Given the sluggish rate at which Scotland has reformed its land use, will we live to see actual predation of mountain hares? I'm not so sure.

But we mustn't underestimate mountain hares as a species either. 'We know from other studies that evolution can rapidly take effect when the need arises,' said Scott. Take the peppered moth, for instance. When industrial pollution killed the lichen off many trees, the moth's rare dark morph suddenly became well-adapted, more camouflaged against these new, sooty backdrops. Within seven years, 98 per cent of Manchester's population of peppered moths were black.

'And you know,' Scott added, 'mountain hares have been around for tens of thousands of years – with huge climatic changes to deal with during that time – so they must be fairly resilient.' Indeed they must, because photographer Andy has noticed in the past six years that the hares in the Monadhliath Mountains had become greyer, more mottled than the 'Daz-white' ones free of grouse moors. Although it's not clear yet whether the winter grey coat has a genetic basis, if greyer hares end up surviving, then we might see a genetically darker, more durable local population emerge. I hope so.

I had mixed emotions at this point. On the one hand, I fear that the colour error in mountain hares runs much deeper (and is potentially more gangrenous) than a surface wound, but on the other, this all offers a constructive wake-up call. Mountain hares aren't necessarily losing out. 'For me, it's a chance to promote an awareness of climate change,' smiled Marketa. 'I'm not losing hope. I don't think this study or other work means that those species are doomed – it's important to consider that – but we must put it out there as an example of what could go wrong if we don't act to mitigate climate change.' *Yesss.*

Dr Isla Hodgson works at the University of Stirling and is a leading expert in Scotland's human/wildlife conflict management. Her years of working directly with gamekeepers on conflict resolution have uncovered an increasing desire for conservation. 'Conflicts between gamekeepers and conservationists are getting increasingly entrenched,' she told me, 'but one shared value is a respect for the land and its wildlife, including the mountain hare. That must be our starting point for a more constructive dialogue.' Over the years, there have been too many reasons to do nothing – too many opportunities for delay. 'But we don't have time for it,' Isla added quickly, 'climate change is right there at our door. Right now. We need to instil a sense of urgency in the day-to-day, but in a way that doesn't fuel tension.'

Hopeful change is happening. March 2021 saw the Scottish government announce a £250 million fund to restore bruised peatlands over 10 years. Alongside this came a licensing scheme for grouse moorland burning amid whispers of banning this practice for good. A vision to expand national tree cover across the country by 21 per cent is the fund's goal for 2032. If you showed a picture of a mountain hare in its winter white to a random Brit on the street, I reckon there's a good chance they would recognise it. Public knowledge of mountain hares has grown, and we need to be encouraged by this. After all, the hare is so blended with British folklore, religion, land and legend. Again, I genuinely believe that we love hares – hares that face a slew of environmental stressors, some humanmade, others as raw as rain. Hares are showing us what's *happening.*

The cloak is off, and we can finally see climate change for what it is and pinch the blueprint of its Grand Plan if we get in there quickly. Whether it knows it or not, our mountain hare is serving an era and has become a totem for change. I find it unbelievably cool, but I'm also a frightened little mole.

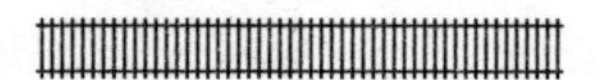

I concluded my chronicles with the mountain hare with some takeaway pizza washed down with a beer as I looked across a darkening Loch Morlich. Water as black as ink softly lapped the shore. White mountaintops flushed salmon pink, their summits reflecting intense, isolated spots of sun. It looked like stage lighting. Every dip, rise and fall of the ridges were naked in this light. I was enjoying a private show. Chilly air descended a breathless sky. A few swigs of beer in, and I began to question whether animals have conscious thought.

Further swigs had me asking a nearby tree whether it knows why the seasons are getting all messed up. Maybe our mountain hare knows. By the end of the bottle, I squinted my eyes tight shut as though capturing the scene, already missing it. Agh. This always happens.

Seven hares and two golden eagles were an excellent insurance policy against the next trip, though, because the following month, I was to stray off at a new tangent, looking for a bird that tries its best to evade all observation and generally hates people. Ideal. Wish me luck.

CHAPTER NINE

Merlin

'Hi, yeah, I'm on the train? Argh – I'm going into a tunnel … hang on –'. It was mid-April. As I waded into an erratic phone call with Mum, I was heading north again. Sheffield, South Yorkshire's metropolitan polestar, was my checkpoint before a 26-mile spin on the Trans Pennine Cycle Trail to a town bordering the Peak District National Park. I felt like an old hand now – a public transport veteran and proud of it! Grey herons had become a surprisingly reliable sight from the train window, bagging me regular spotter points.

Birmingham New Street Station gleamed urban and futuristic through my window. Reflections warped and ballooned across its enormous stainless-steel panels. Between Tamworth and Derby, I donned a cycling jersey in an attempt to summon the mood for a hot, hard ride across the Peaks. In the spirit of blending in, this jersey was neon pink with neon orange sleeves. And still I wonder why, despite my best intentions, wildlife often eludes me.

Sheffield's tram network and pedestrianised geography provoked a tussle between me and its cycle route signs, of which there were many. Part of Sheffield's revamp into an edgy northern hub has been to make it very bike-friendly. And to be fair, they've done a brilliant job, making much more of an effort for cyclists than most UK cities I've visited. Here's hoping Sheffield has set the trend. But whether it was from hunger, fatigue or both, as far as I could see, the signs to the trail all seemed to contradict one another, and I found myself looping circles around a giant Primark, KFC and Wilko, dodging traffic cones and roadworks.

Portions of the Trans Pennine Trail and National Cycle Routes 6 and 67 would pull me where I needed to go, and the elevation profile confirmed a cut-glass shape of ascents and descents. No choice but to get on with it. Rucksack laden, massive and heavily provisioned as always. I had hills to climb and a bird to find.

November 2019 saw near-biblical rains lash Sheffield's city centre, leaving shoppers stranded and roads submerged as the River Don breached its banks. Meadowhall, the retail hotspot of the region, was particularly badly hit. Subsequent storms intensified the fallout, and as I meandered along the banks of the Don, the diversity and sheer volume of litter, plastic and debris in and around the flow were overwhelming. With local authorities still reeling from devastating floods, parts of the cycle path closest to the river remained cordoned off and re-routed. A surprise lump of riverbed had been vomited onto the path, nearly throwing me off my bike, as a descent ended abruptly in piles of filthy sand. Skidding around and back up onto the road, litter dominated the view. It had been this way for a while. Flimsy strands of stringy bin bags draped over branches and choked banks, catching the wind like streamers after some parade.

Feral pigeons sprung into a frenzy as I tried to stuff some cider cans and crisp packets into my rucksack. I soon had to give up – there was just too much. Towels, a fridge, even a mattress had all ended up in hedges and on the banksides at such impossible angles that you couldn't help but stare in amazement. The river did well to chart routes around it all, but its course was interrupted by objects that shouldn't have been there.

Flooding events cause mayhem, especially in a city like Sheffield that hugs a large river. Clean-up operations are often lengthy, complex and expensive. In South Yorkshire alone, the latest flood defence scheme cost upwards of £12 million. I cannot even begin to imagine the distress flooding has triggered within affected communities here. Unfortunately, however, widespread conscious littering appears to be symptomatic of a bigger problem across the British Isles.

Recent psychological studies have shown that the more rubbish persists in an area, the more people tolerate it. During 2020 especially, the threat of Covid-19 plunged most of us into denial and polarity. Sure, many of us sought the healing power of a leafy park, but the physical *condition* of it was likely the last thing on our minds.

I was pleased to leave Meadowhall and Sheffield and continue towards the moors, but the litter I encountered on most of the cycle trail was astonishing. It felt like the aftermath of an illegal rave that had gone on for weeks. But later, I discovered a wonderful thing. Sheffield Litter Pickers is a volunteer collective with a mission to outnumber those who are littering their communities. Nearly 3,000 people gather every week in Sheffield to respond to this growing problem alongside the city council. Groups are coordinated across social media, showcasing photos of smiling people next to bin bags, lifting litter pickers high to the sky.

A few weeks before this trip, I phoned a lady called Sara, whose B&B was ideally located at the foot of the Pennines. I always find Yorkshire accents soothing, don't you? Hearing a buttery Yorkshire voice makes me feel like I'm about to be ushered in to partake of a warm brew and some moist cake. England in April 2021 was still navigating its way towards easing lockdown during the vaccination roll-out, and finding accommodation of any kind was a bit of a task.

The market town of Penistone was my destination, and, posing as 'Most Agreeable Guest', I made the immortal mistake of emphasising the wrong half of 'Penistone' over the phone. Take my advice: if you ever make it to this beautiful part of the world, observe the correct pronunciation – it's '*Pennis*ton' – not 'Penis-stone', OK? Cool. Sitting at well over 200 metres above sea level, Penistone is England's highest market town, and my legs certainly felt this for most of the ride, which was one of many contrasts. The urban blended with the suburban and the rural, swirling together as milk does with tea. Leaving Sheffield's confusion behind, I got a distinct feeling of early Yorkshire. I passed through Ecclesfield and Chapeltown, followed by a few miles of high, open road outlined by dry-stone walls and tremendous views. The distant city sprawl gleamed like pale lakes. It was a good ride.

However, one recurring theme was the enormous potholes, which I avoided with terrifyingly close swerves. After manoeuvring my way around many of them, an unexpected (or perhaps by now to be expected) navigational error had me bumping up the wrong track through Greno Woods Nature Reserve, a green triangle of ancient woodland managed by Sheffield and Rotherham Wildlife Trust. Although the Trans Pennine Trail passes through the woods, I later discovered I had gate-crashed part of the Enchanted Forest Trail, a designated footpath. Occasional bumps were soon upgraded into a full-on rock garden, forcing me to dismount.

As I trudged towards a road I had spotted on my map, breathing heavily and praying I didn't have a puncture, a jay flew across my handlebars in a flash of blue and white. Non-native conifers shaded the forest floor in a hangover from the 1950s. It was quiet, and nobody was about. A chiffchaff jostled its song up the playlist. I felt something brush my left forearm and realised I had been dragging a massive spiderweb along for the ride. God knows for how long. Humming city streets had long since faded. The forest ushered me into the fold. An enchanted trail, indeed. The final swoop into Penistone was a glorious few miles of flat, disused railway, around quiet farms and more woodland. Many people were out enjoying themselves, and it was genuinely lovely to see. My eyes streamed in the wind as they guided me to the B&B. With aching shoulders and tired legs, I emerged from the trail looking like I had delivered about 20 calves in a remote field.

Penistone is strategically parked between the urban triangle of Sheffield, Manchester and Leeds, with 555 square miles of Peak District National Park surrounding it as a buffer. The Peak District's vast moors are within an hour's journey for more than 20 million people, making it one of the busiest green spaces in the UK. Crowned the UK's first National Park in 1951, figures estimate more than 13 million visitors pass through here every year.

Open-access moorland, public footpaths, off-road cycling and disabled access, rolling between limestone dales, farmland, barren moorland and reservoirs is a reasonably accurate summary of the Peaks. Don't be fooled by their name, though: 'peak' stems from an Anglo-Saxon tribe thought to have settled here rather than any actual mountains. Yet even the most experienced walkers can run into trouble on these moors where the weather can change as swiftly as a mood, so maintaining a foundation of respect for the elements up here is always a good shout.

A multitude of wildlife calls these lands home, including more than 160 'priority species' for the UK Biodiversity Action Plan (BAP), the futures of which are all a little delicate. The Peak District pie comprises three contrasting regions. White Peak is perhaps the favourite – an undulating limestone plateau with dales, limestone rivers, meadows and grassland supporting thousands of insects and birds – wholesome. Then South West Peak is a bit of a patchwork, with small sections of moorland interspersed with pasture, hedges, that sort of stuff. Finally, the Dark Peak is the lonely, barren land of vast gritstone plateau, peat and acid grassland. A bit of a wild land if you like, but under the proviso of there being more human than beast – at least in some parts. It was to head towards this lonely, barren land that I awoke at 5.30 a.m. folded inside my duvet like a samosa. My friend Jack lives locally, and he had gallantly agreed to my lassoing of his help. He would be arriving soon. Far from the Saturday ideal, we were embarking on a 12-hour traipse across desolate Dark Peak, hoping to coax the UK's smallest falcon into the skies.

You may as well strap yourselves in for a bit because I'm about to try and fix merlin firmly in your mind. For those of you (I'm guessing, a minority) who already know this fierce little bird, you'll appreciate the task ahead. The other day I typed the word 'merlin' into Google, and it became a sad little game of hide and seek. I tried to find any reference to birds in between Merlin Entertainments (with the world's *First! Standalone! Peppa! Pig! Theme! Park!*), and the 2008 TV series *Merlin* (a thrilling drama of the teenage woes of young King Arthur and Merlin the wizard – televisual gold for me and my brother, Tom.) What a time to be alive.

Now, I know what you're thinking. *Which came first, though? The wizard or the bird?* Some friends and I exhausted this puzzle over a few drinks before I left for the Peaks, hypothesising an array of grand, mystic ideals where writer T. H. White invented the bird itself. Slightly dull conclusion, however – their shared name is merely coincidental. Nowhere on the first page of web searches for 'merlin' are we told it is a bird, let alone the UK's smallest falcon and probably our rarest. Nothing. Nada. Only at the very bottom of page two does its life finally surface – two sentences sitting atop a picture of TV's *Merlin*, actor Colin Morgan.

Further into my investigation, I realised the merlin's sheer famine of celebrity also extends to the research department. Various scientific papers from the 1980s and 90s detail observations about merlins. A handful of reasonably up-to-date studies followed those, but friends, it's been an absolute *mare* trying to find out much about them. During a particularly low week, I even considered ditching the merlin to write about another climate-threatened species. I should be ashamed for such near-betrayal. Even the experts I will introduce you to shortly acknowledged that the merlin tends to hide beneath people's radars. Compared with His Royal Highness, the Peregrine Falcon or Sir Hen Harrier, the merlin is a bit of a mystery.

I was amused (annoyed) because nearly everyone I spoke to about the merlin asked, 'sorry, can I just ask *why* you're writing about merlins? I'm not sure they're an obvious choice.' Case in point, Your Honour. Not one to shy away from a dare, I decided that I liked this about merlin. Brits love an underdog. And, as you'll come to appreciate, so do those who know the merlin: *Falco columbarius* – the tiny, silent assassin of British uplands. I dare you to love it.

Before we get mad for merlins, allow me to zoom us out to a wider view for a moment. I've been trying to pin down what our problem is with birds of prey. We seem to have a real issue with them. Birds of prey are collectively described as 'raptors', which stems from the Latin *rapere* meaning to 'seize' or 'capture'. Our merlin is one of these. But here comes the sauce: we've simultaneously idolised and persecuted raptors for centuries. And thinking about birds generally, UK skies host 40 million fewer birds than they did only 30 years ago. That's a rapid exit! Forty million birds gone – within my lifetime. Such colossal declines are hard to dissect. Yet what is easy to see is that our relationship with birds of prey needs an aggressive makeover if we are to avoid further losses. Makeover scenes are always the best part of the film, right? And the good news is that this is one of the simpler things we can change, and we can start right now. Seeing a bird soaring above you is a good part of your day, is it not? Well, making that happen as much as possible is a rare and straightforward case of us truly having the power. But we've sprung ourselves a trap by failing to commit to an attitude. Big mistake.

Take eagles, for instance, of which the UK is lucky enough to have two species – the golden eagle and the white-tailed eagle. Held in worldwide regard as symbols of freedom and power, eagles are a timely reminder (for some) that humans are indeed connected to the divine, so high do we soar on our little attempt to stay at the top. The idea that an eagle is an all-knowing, all-seeing oracle was a key angle of ancient Greek mythology. So much so that Zeus – the bearded god of the sky – enlisted an eagle as his divine messenger and was rumoured to have transformed himself into an eagle when he fancied it. I fancy it.

To appreciate that the eagle has established itself as a pretty global asset, one only has to consider the Aztecs, the Native Americans, the US dollar, and Barclays Bank. Eagles and other raptors are also often among the answers to those online quizzes like 'What's my spirit animal?' I've had a go. In one quiz, I am a kitten, and in another, I am a snake. My point is that time and again, across literature, folklore and society, we have selected hawks and their cousins to represent our most treasured ideals – leadership, strength, spiritual awareness, courage, progress, transformation – to the point where every single bird of prey today embodies a living paradox. We've borrowed or even stripped these birds of their most iconic attributes, cloaked ourselves in the values we've ascribed to them while leaving them exposed and fragile. How very rude of us.

By now, you've probably noticed our track record of damaging the things we don't understand. It's ageing poorly. Over the years, killing raptors has become established as a bit of a hobby (as it were). Records from the Edwardian and Victorian eras (though I fear the timeline stretches way back) would have us believe that predators existed to be killed, and killing is what thou shalt do, Your Grace! At some point during this period, governments were placing bounties on the heads of certain raptor species. For instance, the red kite, a legally protected and cherished street-cleaner during the Middle Ages (scavenging dead things), entered the sixteenth century as 'vermin' with a hefty bounty on its head. They killed them because they were a bit too big for their liking and thus a certified 'menace'. Yet the devastating truth is that a red kite could barely kill a frog, so weak are its legs and feet. Centuries later, populations are still recovering, but only through intervention and reintroductions. There are parts of the UK to which this species has yet to return.

It wasn't just a British thing either. Raptor persecution was common worldwide – confirmation of the raging storm humans cause when they're at the helm. Shootings. Poisonings. Trappings. The hen harrier is still the UK's most persecuted bird of prey, despite being legally protected. Some conservationists risk their lives in pursuit of justice for Britain's raptors. Bird crime is a grim truth.

One more snack for thought. Whether humans have evolved to thrive on slaughter and violence continues to divide anthropologists. We may possess some biological imperative for violence that *we* instigate, but we cannot bear to know when wildlife does the same. It frightens us, and that's when we lash out at the natural world. For we want nature to be adorable, cute, cuddly – a comfort! We find the very thought that an animal might need to rip another to shreds to feed its young most distressing. A friend told me recently about an exchange on Twitter. Someone posted a picture of a ring-necked parakeet (not native) ripped apart by a peregrine falcon (native). Then some salad informed the rest of the Twittersphere that he would unfollow the person who posted it for promoting such horrific content. Raptors are carnivores. Raptors eat other animals. Nature is brutal. Nature is sinister and indelicate. It wants to survive too. Can we please just get over it?

I'm told that people like a merlin's legs, that they have a bit of a thing for them. I think it's fine to objectify nature. Large, yellow and flashy, merlin legs are of a similar vintage to the black guillemot's red pins, but the merlin wouldn't care less about this and most likely would have a tystie for dinner. Bless you, tystie. I was thinking about all this while I was waiting outside Penistone train station for Jack. You may know him because, aside from being a talented naturalist and podcast host, Jack Baddams' videos of long-tailed tits have catapulted him to mild social media fame, especially among a fan base that South Americans surprisingly dominate. Jack grew up in 'the North', and I trusted his birding wisdom to deliver me some Saturday merlin.

However, I knew our chances were slim, and various colleagues warned me to lower, if not quash, my expectations. Merlins are the kind of bird that even birders forget about until they see one. They are British residents. Migrant visitors that often breed in much chillier northern European latitudes boost our winter population. For visiting merlins, the UK lies towards the southern edge of their range, and so they prefer sticking around on the British uplands, loving a moor, enjoying a hillside. Our breeding merlins are known to nip down to southern coasts to spend the winter, often joined by others popping south from Iceland.

A devastating mix-up with organochlorine pesticides like DDT[33] plunged the merlin into near extinction in the UK during the 1960s and 70s. Although merlins are making progress, their recovery has been painfully slow. Since the 1960s, people have realised that spreading clouds of chemicals over the land is a tad counter-productive.

Like all predators, merlins have a seat near the top of the food chain. When toxins accumulate among species lower down that chain (like the insects, small birds, and small

[33] Dichlorodiphenyltrichloroethane. Infamous for grim environmental impacts until, its UK ban in 1986. If you want to learn more about this, you must read Rachel Carson's 1962 classic, *Silent Spring*. (And good luck in the Spelling bee.)

mammals that merlins eat), those toxins have stockpiled to lethal concentrations by the time they reach the merlin. Like many of the species we've explored together, this historic hardship has at least afforded room in the filing cabinet for some protective merlin paperwork. Merlins are classed as a Schedule 1 species under the 1981 Wildlife and Countryside Act and the 2004 Nature Conservation (Scotland) Act. Unsurprisingly, any nest disturbance, injury or egg-stealing is a criminal offence.

I was pleased to discover that the merlin has mostly escaped the targeted persecution from which other raptors suffer. One of the main reasons for this is that it is too small to hunt red grouse (yeah, them again!), so it doesn't have cause to rile grouse estate owners in the Peaks and parts of Scotland. Phew. Although conservationists are happier with merlin numbers now, they are still a red list species in the UK. As I write this in 2021, current figures estimate around 900–1,500 breeding pairs. But, like train fares, this status changes every year. Nothing is certain. Standard assessment seems to conclude British merlin populations are relatively stable but vulnerable to local declines. Interestingly, the IUCN (International Union for Conservation of Nature) has popped them on the list of species of 'least concern' globally, showing how important it is to get things right at home.

'Pigeon hawk' is how the Americans describe the merlin. Flattering? I can't decide. Although very similar to the American subspecies, our little Brit is heavier, stockier and hasn't quite conquered the urban niche to the same extent. Please note a merlin's density. I come with a similar warning: *much heavier than I look*. Like Tom Cruise, Her Majesty, the Queen of England or your white blood cells, our smallest falcon demonstrates brilliantly that size really does not matter. Often, the smallest of things are the most tenacious. The mere thought of a merlin offers us wise counsel indeed.

Males have dark slate-grey upper feathers and a brown chest. Slightly larger, females weigh up to 300 grams, with browner outer features and a creamier chest. It also means that the *males* are the UK's smallest falcon. A merlin's black tail band offers a tidy aesthetic. A single white stripe of feathers above each eye fashions a glorious brow. Overall, they're striking little birds. Sexual dimorphism is common in raptors as it opens up a more varied prey selection. Differently sized males and females hunt slightly differently, and combining different predatory techniques works a territory more efficiently. Male merlins are also known as 'the jack'. My Jack for the day had arrived on time, and we drove to a merlin-y region of the Dark Peak where bird ringers have monitored nests for more than 30 years. Wheeling a bike across that part of the moor is not allowed (merlins are ground-nesting birds, and my trip was at the start of their breeding season). And to be honest, as I was slightly saddle-sore from the day before, I was delighted to play enthusiastic passenger for a short drive.

'This ...' Jack's voice trailed off as he glanced beyond the windscreen, '... this is a black hole for birds of prey,' as we dipped down an open, steep hill. 'I love the Peaks, I really do.' he told me, 'Objectively, they're beautiful, you drive through with blissful ignorance, but it's also as artificial as a golf course.' Fair enough. A good chunk of the Dark Peak is privately owned for driven grouse shooting. Like in the Cairngorms, open heather moorland is vital for ground-nesting birds, provoking similar tensions. Jack explained we were nearing the worst 10-kilometre spot for raptor persecution. 'Birds come in, and they just don't go out.' But what's puzzling is that merlins do OK here.

Parking shortly after 7 a.m., we climbed a rough track that widened quickly to reveal our height above the small valleys and reservoirs below. Wind turbines stood across distant hilltops, whirring about half-heartedly, as though they were unsure if they'd turned up to the right event. 'When you're tracking merlin, there's *a lot* of walking,' Jack laughed. Fine by me. The day was looking lush already, and Jack was already regretting his layers of thermals under his outer clothes. Soon I, too, was shining like a disco ball. Grouse were everywhere, laughing at the same joke as their cousins in Scotland only a few weeks before. Dry-stone walls zipped up vast drapes of threadbare moorland. Dry heather crunched loudly at each step. Those swathes that had been routinely burned for grouse management even more so. 'A low diversity of plants means there's barely any seed bank up here,' Jack remarked. 'The land is being pushed to within an inch of its life.'

Sponging up the vast space around us, I was reminded that we were looking for a bird that's about the size of a blackbird, flying a distinctively fast, energetic flight low to the ground in a landscape with a matching colour scheme that will swallow it. I sighed as I also remembered those recent figures and recalled that we were looking for one of just 30 known merlins across this entire Dark Peak desert. Fat chance. But we threw the day at giving it a go.

Years of practice and patience has gifted Jack with the ability to see things that I cannot. A golden plover, for example – a pretty little wader I wasn't sure I had seen before – busy performing its noisy display flight and wheeling above us just like any other bird. I quickly learned to get excited when Jack got excited. To fall quiet when he fell quiet. When Jack's binoculars lifted and scanned fence posts, mine followed suit. Like most raptors, merlins tend to time their nesting a little later than other birds to monopolise as many fledglings from other species as they can in early spring. 'The merlins are not quite settled in their nests yet,' Jack told me, 'so they may well perch on these boundary posts on lookout.' Jack could have told me anything in that moment, and I would have believed it. We had every chance of seeing a merlin. I crossed my fingers tightly at my side.

Something upset the line of heather not far from where we were walking. Large, white and distinctly mammalian. 'I don't bloody believe it ...' I muttered. Jack laughed, 'Ha! There's your hare, alright.' FFS. There it was, lighting up the heather with its arrogant little arse. All that frolicking up to the Cairngorms and the mountain hares were here in England all along. Clearly, this hare was a descendent of the hares translocated from Scotland as game to English uplands during the early 1800s.

We ended up seeing more hares that morning than I ever did on their native slopes. Greyer and more mottled, but they still stood out like little beacons against a haze of heat that already shimmered above the ground. Don't get me wrong, though. Seeing hares felt like bumping into old friends. The swathe of monotone Dark Peak we were hiking across housed the most bizarre assortment of species.

Few physical barriers limiting birdsong and animal movement exist here, which shifted the boundaries of what we could see, smell and hear. Gamekeepers ensure a stark lack of crows and foxes, meaning waders and plovers get more of a say in things. Every cloud, I guess. A curlew, the largest wader in Europe and one of Britain's most-threatened, cried high above us, its calls stretched by the wind. A very wild sort of mood descends when that happens.

The lapwing – another plover (from another mother?) associated with wetlands and farmland across the UK – is also in decline, and yet I saw more of them than I ever had that one day. Merlins have a taste for lapwing, so our eyes were officially peeled. The first time I ever saw a lapwing, it took me a good while to realise that the police car siren that was harshly blaring across the road was a black/white/iridescent bird performing an unbelievable air tattoo above my head. Its scientific name *Vanellus vanellus* means 'little fan' – a nod to its iconic, flappy flight. Its rounded, paddle-like wings allow it to tumble, ripple and corkscrew through the air. A lapwing gives its entire body to the sky, pouring itself through invisible cracks and tunnels. It is dizzying and hypnotic to watch. (Can recommend.) But be prepared for any sort of meaningful moment here to be interrupted by greylag geese – loads of them. Their clumsy honks and squabbles signalled flustered indecision about whether to hang out on the moors today or on the reservoir. 'Should've written a chapter about them instead,' Jack chimed in. Cheers, mate.

Earlier, I told you some conservationists had risked their lives to seek justice for birds of prey – justice in all senses. Enter Dr Ruth Tingay, one of those people who undersell their achievements by describing themselves as a 'conservationist, researcher, blogger'. Friends, she is *so* much more. You may have gathered that tracking down merlins is hard. And tracking down humans who have worked with merlins, let alone finding a woman operating within the 'raptor' space, is a challenge too. But as I was doing some pre-trip research, Ruth's name kept cropping up in various circles: 'Wait, you haven't spoken with Ruth yet? You *must*.'

With a global career spanning five continents, Ruth is a leading voice on raptors and their persecution. She was an international director of the Raptor Research Foundation for six years, followed by four years as its president. Her blog, Raptor Persecution UK, has been influential for its value to science, policy and the law. Ruth has written countless papers and campaigned relentlessly. She has been seriously threatened directly – multiple times – for giving a shit. She was also instrumental in securing the grouse moorland licensing I mentioned in the mountain hare chapter. 'It's a big deal, but it's not over yet,' she admitted over Zoom.

Ruth is risk-ready and considered one of the UK's most important conservationists.

George Orwell told us in *Animal Farm* how, although equality prevails in the animal kingdom, we cannot deny that some species will always rise above others. I believe this also applies to humankind. Ruth stands taller than most people (she'd never agree with me, of course).

'Raptors have just always been better,' she smiled over the screen. She exudes warmth yet is as hard as nails. I'm not sure you'd commit years of service in a battle for raptor rights and not develop some sort of armour. Ruth regularly speaks publicly about a concept known as 'wilful blindness' – a term often batted about in legal settings, describing a conscious avoidance of the truth. Although commonly mentioned in the context of illegal raptor persecution, the concept of wilful blindness applies to British conservation as a whole. 'It's a dangerous mindset,' Ruth said darkly.

Still keen to uncover why our national attitude towards raptors is so polarised, I asked Ruth what they do for her. She laughed, shrugging, 'They completely dominate my world … I still don't really know how to answer that question properly.' A debut rendezvous with a sparrowhawk in Windsor Great Park – 'doing a bit of rhododendron bashing' – during her time as a volunteer sealed her fate with birds of prey. 'I was transfixed. I had no idea we had birds like this in the UK. When I realised a lot of people want to kill them, it became obvious to me that I would study these species.' Not studying for the sheer hell of it, but so her knowledge might contribute to the protection of raptors.

While the merlin hasn't been a direct species of focus for Ruth, as it (so far) largely escapes persecution, she has studied merlin populations in the Western Isles of Scotland and is aware of the merlin's fragility. 'An increase in heather burning on the Scottish moors is stealing the merlin's nesting habitat,' she told me, describing how long, bushy heather – waist height in places – makes for a better quality nest, not only for the merlin but also for its favourite prey like skylarks and meadow pipits. Imagine the turmoil that would ensue upon trudging up to bed, exhausted, only for you to meet with a charred heap? Now imagine you had offspring that had perished along with your organic linens? Grim.

Back to Ruth. 'Merlins *can* nest in old crows' nests, which is smart. It protects the nests from stoats and weasels, and the high aspect affords a view to spot prey.' But, recycled nests are not a reliable alternative any more. Sheltering belts of trees lining the edges of moorland were once commonplace on grouse moors. Crows would roost, then move on, leaving their former refuge up for grabs. Too quickly, though, these corridors have been felled – replaced by swathes of heather seed. And grouse.

Ruth told me about a 30-year-long study of breeding merlins across four grouse-shooting estates in Scotland's Lammermuir Hills, which reported this trend. Over three decades, the researchers measured a decline in merlins and other birds as grouse moorland management intensified. Beginning field observations in 1984 offers an unprecedented window into how these birds have coped with some extraordinary shifts in pressure on the landscape. And while this study focused on Scottish birds, and here I am telling you about my spree in the English uplands, it is all related. This 30-year Scottish dataset provides us with a rare gift of foresight. One only has to look up and down the Lammermuir Hills to note the stark similarity of their tessellated, burnt hillsides to those of the Dark Peak. 'You can't not be concerned,' Ruth observed.

When 10 a.m. in April feels like 2 p.m. in late July, you know something is not right. Poor old Blighty doesn't weather temperature swings quite as elegantly as the Algarve. As with most species we've considered so far, the climate change link to merlins is annoyingly murky. But only deniers would keep their heads in the (potentially increasing) sand. Cambridge University led a study in 2020, which found that recent UK summer droughts are worse than any of the past 2,000 years. Yet on average, each person in the UK blithely uses up to 140 litres of water every day, most of which then runs down the drains, never to be thought of again.

The researchers from Cambridge were able to gather data so deep into the past thanks to a ground-breaking technique. Careful analysis of the carbon and oxygen fingerprints left on European oak trees revealed a startling timeline of the UK summer climate. Dry weather and drought have become less of an anomaly and more of a trend, steadily increasing in frequency and intensity since 2015. The study concludes with strong suggestions that human-induced climate change is the antagonist, alongside extraordinary forest dieback.

Turning the pages back to 2003, research into the record-breaking European heatwave found that human activity more than doubles the risk of such events repeating themselves. At the time, 2003 was the hottest recorded summer since 1500. I have hazy memories of this – of hot, stuffy classrooms and teachers badgering us to drink more water. I think even my parents ventured into the river in their underwear once or twice. Safe to say, it was a year of extremes. According to Oxford University's Environmental Change Institute, there were 2,234 human deaths recorded during this heatwave in the UK alone. The researchers estimated at least 1,117 of these deaths could be directly linked to human-induced climate change. Scientists will need to do much more work to confirm the exact causes of death, but more and more, it's becoming possible to attribute weather-related deaths to climate change. Given our general aversion to 'the death thing', I'm not sure we want to see these data. But if it shocks us into sorting out climate change, so be it.

Mike Price – a local Peak District raptor worker and bird-ringer with a keen interest in all-things-merlin – has done some probing into whether the merlin, as a falcon, should fear the UK's weather rollercoaster – dry or wet. Other raptors are already finding the ride nerve-wracking, and seeing how they're affected can give us an indication of how others might fare. In Spain, nesting close to some springs that flooded proved an utter disaster for some pairs of peregrine falcons. Chicks became saturated and developed dangerously low body temperatures as a result of torrential, unseasonal rains. But as always, with merlins, there's more to it.

Mike admitted the complexity of it all. 'The timing of wet weather seems far more important than the amount,' he said, referring to a wet period during 2019 that affected the breeding success of other species in the Peaks, but didn't affect the merlin. Remember, we are at the southerly edge of a merlin's range (they've evolved to exist within slightly cooler climes), which makes me wonder whether it's these looming droughts and heatwaves that may prove to be a merlin's Achilles heel.

Merlins were the focus of an article in *The Times* in 2015 by Jim Dixon (former CEO of the Peak District National Park, aka 'Peak Chief') entitled 'Hot weather could make the merlin disappear', casually linked their weak numbers in the Peaks with a warming world. Dodging weak conclusions (tricky in science, eurgh), I posed a slew of messy thoughts to Dr Matt Stevens, a conservation biologist at the Hawk Conservancy Trust, a few weeks after my trip. 'It's a difficult one to call and link to raptors,' he wrote back over email, 'most of the species occurring across the UK have ranges that extend beyond our shores in all compass directions and will be exposed to milder, colder, wetter and drier climates. Would this not suggest that the overall populations may be able to cope with the expected temperature increases?' I guess. But everything has its limits.

One other curveball for merlins to dodge is the geography of the Peak District itself – an epicentre of green space for millions of people in northern England's urban centres – 'England's megalopolis' as it's sometimes known. Intense industrialisation during the Victorian era – coal mines, textile mills, railways, canals, shipyards, ports, international trade – made these cities officially *big*. Putting aside the routine burning of heather for grouse for a minute, Ruth Tingay also mentioned how, along with people pressure, incidences of wildfire and arson are mounting on the moors. Five thousand of the 350,000 acres (141,640 hectares) of the Peaks were burnt to a crisp in the spring of 2003. It was a challenge to find out the exact cause. Not only was this a living nightmare for wildlife, but a colossal dislocation of peatlands relinquished carbon from its burial chambers, discharging it back into the atmosphere. Remember, peatlands are our largest natural terrestrial carbon store, and damaging them releases around 6 per cent of global human-induced emissions annually.

Saddleworth Moor was on fire for more than a week in the summer of 2018 – a fire believed to have been started by a group from Manchester seen lighting a bonfire on the Pennines. High winds, unusually hot temperatures and the severe absence of rain that summer prolonged the Saddleworth blaze for days. And in January 2021, five men in Derby were charged with arson after losing control of a campfire while camping near the moors, engulfing a Peak District woodland. In the open countryside, the 'accidental' can so easily spark disaster, and as many parts of the UK experience hotter, drier conditions, that's only going to increase.

The way a merlin moves might interest you. The opening steps of any predator-prey waltz are still largely unknown to most of us, but I reckon you'll enjoy learning about this one. Jack compensated for the lack of anything merlin-related on our walk by telling me everything a merlin *would* be doing if we were to see it. 'Live fast, die young' is a merlin's mantra. 'Hard on them mentally, I suppose, killing something every day,' he mused. Our tiniest falcon lives for around three years on average, has a wingspan of 50–62 centimetres, and is an absolute unit of a hunter. Any human encounter with a merlin is best recounted in a superlative firework, so they are typically described as 'thrilling', 'terrifying', 'furious' and 'exciting'. It is pretty much the only way to depict how they move. Exploding from their lookout post, a merlin hunts with immense purpose and hustle – rarely will it glide like a buzzard or hover like a kestrel. Instead, the merlin uses fast, powerful wing beats, interspersed with short, precise glides, to help it stay fiercely loyal to a low airborne course while it scans for prey. As a rule, all hawks have incredible vision. A kestrel can spot a dor beetle 50 metres away. It (and many other species, too) can detect ultraviolet light, invisible to the naked human eye, affording it a window into the urine trails of small mammals. Handy, if a vole is their snack of choice! Merlins hold their wings close to their bodies in flight, often hunting at less than a metre above the ground. They capture most of their prey in the air, and alarmed birds on the ground are 'tail chased'. If the catch is successful, they will carry their prey to a 'plucking post' – the top of a fence, a stone wall or heathery hummock – where it will be meticulously plucked (if feathery). Breeding pairs often work together, but during early spring, the male is the more frequent hunter of the pair, delivering a plucked, beheaded dinner back to the nest to feed the female while she incubates the eggs.

Some birders claim a merlin's aeronautic agility while hunting can outmatch its larger hawk counterparts. Watch a video of them doing their thing and you can't help but applaud their sheer versatility. Their hooked bills with tiny serrated edges assist the talons at the end of their sunshine legs as they rip open their meal. When a breeding pair hunts together, one may flush a terrified meadow pipit towards its mate, which makes the capture. Because those who flush together stay together.

During our call, Ruth Tingay reminisced about her time studying merlins on the Isle of Lewis, off the west coast of Scotland, between 2003–2005. I gather the Island has quite the winter to endure. 'Fucking *freezing* – you'd have twenty layers, six hats and a sleeping bag while sweeping the horizon every minute looking for them.' For Ruth, the drama of witnessing a day in the life of a merlin made it bearable. 'As soon as anything came into their airspace, the merlins would be up like a shot,' she laughed, demonstrating rapid swoops with her arms over the camera, 'literally vertically, shouting and screaming, buzzing at whatever species was there, be it raven or golden eagle.'

A merlin's motto is simple: 'Don't even start with me, because I *WILL* have a go.' Ruth said watching a merlin buzzing a golden eagle in this way was like watching a mosquito nipping at the eagle's heels, 'dive-bombing, mobbing, in a ferocious, funny kind of way. You have to hand it to them – they're bloody feisty!'

I wanted these anecdotes. I wanted my own merlin story. As I wandered the Peaks with Jack, he told me about merlins' fearlessness around bigger raptors. He said to keep an eye out for a buzzard or a crow because the likelihood is they've spotted something to eat, and a merlin may come along and start mobbing them, driving them off their territory – 'they literally don't give a shit', he told me. Hearing this reminded me of a now-famous photograph taken in July 2020 on the Dark Peak by Steve Gantlett. It shows a visiting (*massive*) bearded vulture[34] being hounded by a merlin. Look it up and you'll see how perfect Ruth's mosquito analogy is.

During my campaign to *Find Out Anything About this Bird*, I discovered that noble folk also appreciated the fun in a merlin's flight. I haven't mentioned 'falconry' yet, the art of using a trained raptor to pounce on unsuspecting wild food or 'quarry', or simply just keeping said birds at your leisure. As Helen McDonald so beautifully recounted in *H is for Hawk*, falconry has experienced a bit of a renaissance. Thought to have originated somewhere in the Far East, falconry made its way to British shores around AD 860, sparking a person-plus-hawk situation that has stayed in vogue through the ages. In 1066, William the Conqueror, heady from his Hastings win, assumed the throne as King of England. Norman castles pocked English landscapes, and with them came intense privatisation. It was then that falconry shifted from an everyman's pastime into an elitist sport. Until then, anyone could train a bird, but soon only the privileged – Henry, Son of Kingsley, Son of Bertram, Son of God – could qualify to 'hawk'.

The 1468 *Book of Saint Albans* ranked British raptors into a social hierarchy. Society had truly embraced social delineation by then, so why not also apply that to birds?

[34] One of Europe's most impressive and rarest birds graced Yorkshire's skies from her home in the French Alps, grabbing headlines around the world. With a wingspan of more than 2.5 metres, the bearded vulture is bigger than the white-tailed eagle. Unforgettable.

For instance, a female peregrine, the world's fastest animal, is dubbed a 'prince' in the book. The male, an 'earl'. The hobby[35], a 'young man'. A sparrowhawk is doing very well, classed as both 'priest' and 'holy water clerk'. But what of our merlin? Mercifully, The *Book of Saint Albans* ranked the merlin a 'lady'. It contains a trio of essays on hunting, hawking and heraldry, and it's probably fair to assume that this publication was written with a more gentrified reader in mind than I can claim to be.

Further reading led me to discover that a female merlin (m'lady) was deemed a noblewoman's most fashionable falconry accessory. How chic! Merlins were flown in stunning displays by ladies of the court, who were more able to handle their small, compact size. The merlin was the preferred bird of prey for Mary Queen of Scots, who enjoyed hawking them at unsuspecting skylarks – no doubt a diversion from tumultuous sixteenth-century life. Some records relate that she continued to enjoy this pastime during her infamous periods of imprisonment.

Back to the Peaks. Jack and I found a beat on our quest for a merlin, with periods of slow walking and talking alternating with silent vigils over an outlook, scanning different portions of sky. Squinting in the sun, we debated at length about what animal we would be at a dinner party, concluding that a herring gull and raven would both have good stories, eat anything and have the ovaries to tell any annoying diners to shut up. We mused the ridiculousness of a sweet wren having the throaty, smoker's croak of a raven. We wondered whether a merlin might be watching

[35] Not, as you may first think, knitting, cycling, gaming or baking, but rather, another small and exciting British falcon – conservation status: green.

us from a perch, mocking our feeble attempts at tracking it. Imaginations ran easily out here.

Thanks to local raptor workers like Mike Price, merlins have been monitored in the Peak District for a long time, although some debate remains as to how well they're actually doing. A thread on Twitter in April 2021 disputed the context around which merlin breeding success is reported. Take, for instance, a retort from our friend Ruth who, as you may gather by now, knows what she's talking about: 'Yes, approximately 50 merlin fledglings from 15 pairs in 2020, but that is well below the target set by the Bird of Prey Initiative (37 pairs), and in an area of 196 square miles of moorland, 15 nests is pathetic.'

Dr Alex Lees, a senior lecturer in biodiversity at Manchester Metropolitan University, followed suit by citing data from the report Ruth was quoting from. Lees wrote that claiming merlins' success on driven grouse moors risks being 'fast and loose with the truth', explaining how their nests' 'repeatedly fail' across some parts of the Peak District. 'From my desk, I can see a driven grouse moor where there are currently a pair of merlins nesting,' Alex messaged back when I probed further. 'But just south-east of me on moors near Glossop, territories have a high failure rate and some traditional sites are unoccupied.'

I find this particularly interesting given that there have been repeated confirmed cases of raptor persecution in the area around Glossop. An osprey was found dead in a sprung trap in 2015, and in May 2020, a buzzard was found that had been shot twice. Alex concluded that merlins do well on driven grouse moors '*if* they are left alone. But that isn't the case in many places.' Indeed, Mike Price explained that merlin abundance here peaked in the 1990s and has been in slow decline ever since – saying that they have been, 'just about holding on for some time'. He told me that the birds appear to be shifting north. The local Raptor Monitoring

Group has recorded nests further north. 'But that is not the full story,' he said because that 30-year Scottish study I mentioned earlier told us that reduced merlin numbers were more often related to pressures the birds faced on the breeding grounds themselves rather than on any external factors (like survival over winter).

Anyway, let's forget driven grouse moors for a bit (happily!) because, as Mike tells us, there's more to it. Quite a lot more. The availability of good nest sites for them and their favourite prey are also crucial. All the experts I spoke to about merlins strongly advised me to go down 'the food chain route'. To investigate that, we need to get up close and personal with heather.

The dramatic-sounding notion that the UK supposedly holds '75 per cent of the world's upland heather habitat' is batted across the internet from time to time. There is no doubt that between Great Britain and Ireland, we can be confident of claiming a very generous slice of the global heather pie, but accurate figures remain vastly inconclusive. A healthy heather upland or woodland understorey is far from a purple monoculture. Wet and dry heath, blanket bog, peat bogs and mosses have meshed together over 5,000 years. Not only does this act as a vast natural reservoir, feeding fresh water into tributaries and rivers and housing the nests and nurseries of countless species, but heather has served humans well, too. The Vikings concocted a prized heather ale, finding heather's twiggy stems to make excellent thatch, broom bristles and mattresses. If we so please, we can soothe stressed nerves, tickly coughs, rheumatism and arthritis by drinking heather tea. Today, we probably aren't making heather mattresses, but we do enjoy high-quality honey, thanks to the rich nectar of our heather flowers.

Taking another loop around our barren portion of the Dark Peak had Jack declaring that 'some birch would be nice, or just a couple of oaks?' The thing about birds of prey is that they will only show up if the species 'below them' are also in good health. Raptors, including merlins, are an exciting thumbs up that an ecosystem is functioning as it should be. Like sharks, whales, lions and leopards, the presence of top predators brings a sigh of relief. A simple, 'Don't worry! The web is intact. We are resistant to the enemy!' So the fact that we didn't see any, and I heard of no sightings reported leading up to my visit, was unsettling. Because as lovely as the lapwing, curlew and meadow pipit are, the overall biodiversity in the Peak District could be *so* much more.

Remember, the rich geology underlying the Peak District has the potential to support exemplary natural wealth. But the failure for such wealth to be realised is what experts fear will hammer our merlin. We know that they can fashion old crows' nests to suit their own, and some research has found merlins nesting in bracken, suggesting adaptive flexibility, or plasticity, to a changing habitat. But what of their prey? Are they as resourceful? As resilient? Some figures estimate that a single merlin will eat up to 900 small birds a year. Such fuel must be essential in powering their launch and fiery orbit across the moors.

One of the most thorough investigations into the Peak District's merlins was published by Bird Study in 2009, led by author Ian Newton. In Newton's study, around half of the observed hunted birds seemed to be pipits. Two other comparative studies that followed noted a disproportionate preference for meadow pipits in a merlin's diet. A flurry of terrified pipits suddenly erupting from a bank of heather could, therefore, indicate a merlin is in hot pursuit. These pipits are small brown, drab little things, and their sweet song is the most common across the uplands. Yet overall, UK numbers have been in a steady decline since the mid-1970s, placing them on the amber list of species of conservation concern as of writing this book.

Meadow pipits are strongly associated with heather moorland and underlying acid grassland, and they rely on a rich invertebrate menu for survival. They're not the only ones, of course, for skylarks – one of the most endangered upland icons – need to rely on invertebrates too. Whenever I hear the chaos of a skylark's tune, I'm thrown back to waking up at dawn on Dartmoor a few years ago following a summer solstice bivvy with my dad. A skylark sang me out of a dream. Trying to dissect the countless lyrics in its song was a tall order at 5 a.m., but it's hard to forget this bird's victorious ascent high above my sleeping bag against a lilac sky. These are the moors I want for the Peaks.

Survival expert and official Channel 4 hard man Bear Grylls regularly reminds humans what we need to survive: food, shelter and water. Simple. Food and water security are always a top priority in the UK Parliament's various revisions of its Agricultural Bill. Yet, the public health issue of malnutrition continues to cost the NHS nearly £20 billion per year in England alone.

It occurs to me that the inhabitants of our uplands could file similar reports. Crane flies – or 'daddy longlegs' – are upland dietary staples for meadow pipits, skylarks, golden plover, stonechats and many other songbirds. As #basic as bread. Their long spindly legs and dense, slender bodies make for a hearty snack. Part of the 'true flies' group of insects (as is a housefly), crane flies are among the most ancient of flies. When I typed 'crane flies' into Google, it asked me if I meant to write, 'should I kill crane flies?' But whatever you think of them, keep in mind that it's only when tiny herbivores like crane flies are farming our undergrowth that the rest of nature can truly begin.

Thinking about it, what alarms me is how defenceless crane flies are in the face of drought. Their larvae, known as 'leatherjackets', fizzle and dry (desiccate) when temperatures are too high and soil moisture is too low. (Perhaps they look like tiny rolls of leather when this happens?) Data from a PhD thesis from the University of York in 2012 elegantly showed how adult crane fly numbers increase with soil moisture and decline alongside a falling summer water table. Data aside, all you need to worry about is that upland bird survival directly depends on that of insects. With drought exerting a growing annual influence across the UK, this desertification of the supplies that feed the vital links in the food chain poses a severe threat.

'It's already happening,' Ruth Tingay stressed. 'Changes in temperature are affecting insect populations, which is knocking on to the pipits, the skylarks … the merlin.' But we know this. We've played this domino game before. Warmer springs are causing woodland birds like great tits and pied flycatchers to miss the peak of caterpillar prey to feed their young, because the caterpillars have emerged earlier. Under the water, we are also seeing this kind of de-coupling of seemingly indestructible predator-prey relationships between zooplankton (the equivalent of insects in the sea) and fish. As always, the top predators will eventually lose out. *Everything* will lose out; it's just that the top predators get blasted by the accumulation of hurt up the chain. Buses continue to be missed, leaving isolated passengers stranded. With all the issues it's facing: drought, wildfire, potential persecution, starvation … maybe the merlin should quit while it can.

It's easy to get lost in all this. I get it. Despite its tiny size, the problems facing our smallest falcon feel enormous, don't they? But it's reassuring to hear researchers asking questions too. 'My concern is that we won't notice a significant change and will continue with business as usual – while the changes gather even more momentum,' said Matt Stevens from the Hawk Conservancy Trust. Once they had tried to talk me out of writing about the merlin, all the people I spoke with stressed the simplicity of the main concerns facing merlin populations. 'We're not learning fast enough.' 'What if nothing changes?' 'The government isn't listening.' 'If we know what to do, then why aren't we doing it?'

Further studies investigating crane fly vulnerability to drought, for instance, demonstrate how simple peatland management strategies can quickly switch their fortunes. Blocking drains can return moisture to the soils. Restoring water retention republishes the story of heathland as a vast natural reservoir, preventing the desiccation of those insect larvae and bolstering the food and water security of upland birds for pipits and merlins alike. May 2021 saw new legislation banning peatland burning across protected areas of England. However, it wasn't welcomed by everyone, with many conservationists challenging its lack of clarity on what peatland is and what protection it needs.

Perhaps it may help to imagine all this as though we are fitting our uplands with bulletproof vests, tucking a fire blanket around them, and backing up their aeons worth of data to the cloud. Whatever analogy works for you, deceiving ourselves has to stop. We've got to *do* something.

Overlooking birds of prey – be they peregrine, osprey or merlin – should startle us. Birds of prey aren't going to solve climate change by any means, but they can tell us about our relationship with nature. If we leave them to get on with their lives, they will more than likely bounce back during difficult times. Talking of which, a few days ago, Mike sent me an email with a photograph of three merlin chicks huddled in a fuzzy feather ball in their nest. I stared at it for a long time. I zoomed right in on it until feathers filled my screen. In a world that's increasingly terrifying to predict, we should embrace merlins fiercely.

My day on the Peaks was drawing to a close, and this time the merlin stayed true to itself. If one did escape our gaze, we'd never know. Our merlin continues to elude. And I think I like it that way. Ruth likes it too. She wrote back to me after I got home, and I love her response:

Sorry you didn't see any merlins...but secretly pleased, because a big part of the enjoyment of this species, for me at least, is having to work hard to find them! They'll be all the more enjoyable when you DO see them, and I have no doubt that you will.

R x

On the train home, for the first time that day, I caught sight of my reflection as we went through a tunnel. Half of the Peak District's heather was decorating my hair. I looked like a sunburnt medieval washerwoman with a tinny G&T and an iPhone. *Wow, get it together, Soph,* I thought. After all, who knows when one might get the chance to dress up and partake in a spot of hawking with a merlin like a proper *lady*? Finally, back in Exeter, I rode along the River Exe in a trance, and it was past midnight by the time I crept through the front door. I had barely been away more than a day and a half, but being home again felt so good. I needed spring to advance a few more weeks before I crossed that threshold once again. A bee was about to wake up from hibernation, and I wanted to greet it. Come on, then.

CHAPTER TEN
Bilberry Bumblebee

You may have noticed that the Great British public adores giving directions. When given a chance, we come into our own, tossing aside our iconic reserve to guide the long-lost traveller to a far-flung destination. I've overheard textbook iterations of such dialogue during my travels for this book. Almost always, we deliver directions wrapped in a bundle of well-meaning – yet useless – detail. It's very endearing: 'OK, so you know that street? You know – *that* one? Up there? The one you go up before you get to the lamppost with the sign on it? Next to the fence by the bin? Well, turn right there, then carry on past where that field used to be, and it'll be on your left.'

This poor woman. I stood behind her while queuing in the Co-op, collecting provisions for my ride the following day. Her face remained utterly blank as the shop assistant mangled instructions to the nearest chemist. The preceding couple of weeks had plunged me into a weird mood. I was restless and distracted. Returning from the Peak District meant one final trip remained in this little saga. One more ride. I know this isn't meant to be a box-ticking exercise, but I desperately hoped that the final animal on my mind would let me in. Merlins were still darting around my subconscious and no doubt will continue to do so. But, unlike the chemist lady in the Co-op, I was a local and familiar with the terrain ahead, feeling lucky that the rare pollinator I wanted to see shared my county.

Since my family moved to England in 1998, I have been invading Dartmoor's personal space fairly regularly. The outskirts of its 950 square kilometres of national park lie just over 10 miles from home and have been a steadfast persuader, luring me and nearly 8 million others each year. I love that this iconic national park is within reach – that its moors are there when I need them. But to start with, my brother and I couldn't understand the concept of walking for *pleasure.* Although very young, our little American heads had already identified that a car just made more sense. It didn't matter if the store was a mile away or whether Jack and Mia's barbecue feast was just around the block – have car, will travel. As a family, we often recall the hazy memory of our primary school agony, walking the famous slope to Haytor Rocks on Dartmoor's eastern edge. My brother Tom threw himself on the ground, beating his fists and kicking his frog wellies against the rough, waxy grass, refusing to believe that he had to use his actual legs to climb a hill. I followed suit. 'Is this walk a circle or do we have to go back the way we came, ugh?', '*Whyyyy* are we still w*alkinggg*?' and 'Is there a gift shop?' being my only contributions during early walks. What a waste of bloody time I thought walking was when I knew I could be back home gelling my hair and washing Barbie's. After all, being five years old is very serious.

Thankfully, over time, these uplands have changed us. Something about the expanse, the wildness of it all, has quite literally broadened our horizons. I honestly believe that Dartmoor is responsible for untangling our young minds and planting us on our feet. Adolescence then wove a different pattern. Aside from refusing to acknowledge that my body was changing, hills became a challenge I craved to face. Summits, a satisfaction. Height, a hunger. Few of my friends shared these interests at the time, but I never wished they would, often pretending it was all my little secret, to either be enjoyed alone or with my family. Later, when I was 17, I met a boy on these moors, and I like to think that some part of Dartmoor has glued us together all these years. Later still, I discovered how the muscle of friendship could pull you back onto the map. I've posed big questions to howling gales and red skies, finding parts of myself that I didn't recognise ... parts that I like. Places like Dartmoor have an uncanny ability to strip a person down to their rawest form, I think. A growing approach to counselling called 'wilderness therapy' is built on the idea that time in the wild can cleanse gloomy thoughts. I've also long considered Dartmoor (or any vast open space, to be honest) the ideal setting for an insightful first date. Southern Britain's largest area of open moorland leaves nowhere to hide. Truth prevails.

Time had a different attitude this year, it seemed. How was it already early May? By this point during the previous year, we were comparing suntans over Zoom. Yet suddenly, spring 2021 had decided to pause. Following the drought of April, average temperatures dropped to their lowest for this time of year since 1992. Winter clothes reluctantly reappeared. Sunscreen was side-lined. We were desperate for rain, and thankfully, much was forecast (a little too much, in hindsight). But I needed the rain to hold off for just one more day because the bee I was hoping to find might otherwise stay hidden underground. Sensible, yes, perhaps even wise. And yet highly annoying given the hills I would have to climb to get there. Only a short while before, I realised that somewhere in a hairy little valley lives a colony of one of Britain's rarest bumblebees. They weren't the only population on Dartmoor either, but certainly one of the last.

And so, before we part ways, I have one more species to share with you: the bilberry bumblebee. *Bombus monticola*. Britain's very own 'mountain' bee. Yeah, I know! Last time I checked, Devon didn't have any mountains, let alone a bee that is repeatedly celebrated as possibly the most beautiful in Britain. Oh, ignorant me. Like our marsh fritillary butterfly, this bee navigates the uplands with a solid aesthetic. We'll unpick the detail soon, but for now, know that queen bilberry bees are always larger than their male and worker counterparts, measuring around 15 millimetres long. Males and workers, on the other hand, measure up to 10 millimetres. Breaking free from the yellow-black-yellow-black topcoat flaunted by most bees, bilberries have opted for a warmer hue. Even from a photo, I already loved it. Always a fan of an intense orange, I envied its bold accents of colour. The rear end of its abdomen is a deep, rusty red. A bright yellow collar wraps the front of its thorax. And as well as being one of the most vibrant, bilberries are one of the UK's fluffiest bumblebees, their dense fur coat trending on both the ski slopes and the fashion week runway. They can be easy to confuse with the early bumblebee or the red-tailed[36],

[36]The early bumblebee is a small, widespread bumblebee found across the UK. The red-tailed bumblebee is also widespread and abundant. Both of them have distinctive red tails.

which have similar colouring. But the bilberry bumblebee has this glorious red hemline that covers more than half of its abdomen. So there. Male bilberries also have yellow facial hair, which is a sweet finishing touch.

Also known as the 'blaeberry' bumblebee (by the Scottish), this unique insect is partial to living at altitude in the British uplands. 'Monticola' shares its meaning with the word 'mountaineer', which instantly conjures an image of a furry bee donning a rucksack and hiking poles, perhaps taking a selfie atop a windswept cairn. Anyway, these insects live a harsh existence that (surprise!) humans are making even more challenging. (For a final time, please hold that thought.)

Northern European latitudes have a historically wide distribution of bilberry bumblebees; from mainland Britain to Ireland and Scandinavia's ridgelines, the Alps, the Dolomites, the Pyrenees and the Balkans. Some plasticity can exist. If they already live at a higher latitude, these bees can enjoy a nearby lower altitude. They have been seen buzzing around at sea level in Scotland, for example. So it's unsurprising then that very few have been recorded in the flatter, balmier south-east of England.

Guys, this is a bee after my own heart. I'll also admit that I never really wondered why bumblebees are so furry – never thought to ask. Sure, like so many of us, I like them and understand that they're important, but until recently, my mind hasn't bothered to stretch beyond the shallow end of their 'cuteness'. But what I've learned is that all bumblebees are insects that have evolved to pollinate, sleep, breed and repeat in cold environments. Cooler climes are the preference for our tiny, chilly specialists, and as we'll find out, a warm, furry jacket protects them against the worst of it. Who would have thought?

Of all other insects, bumblebees might just be our favourite, don't you think? Put it this way, 'oh look, a bumblebee!', is exclaimed far more dreamily than if it were a wasp. If I was (cruelly) asked to rank my top 10 insects, bumblebees would predictably make the cut. They may even sit in my top five. Many different types of bees exist across the planet, and it's more than OK if you get them all a bit mixed up. For instance, I only truly understood the significance of solitary bees when asked to write an article about them a few years ago. UK bees can be split into three groups: honeybees, bumblebees and solitary bees. We have one species of domesticated honeybee in the UK, and this is the only bee that makes the honey we eat. Honeybees live in hives of up to 80,000 female workers and males (drones), governed by one fertile queen.

Solitary bees sport the lowest profile, yet they are the most species-diverse, and we have more than 250 in the UK alone. Some are so small that they are mistaken for flies. Severing the apron strings from queen and colony, they favour a solitary life. True specialists, solitary bees have diversified into roles, such as 'leafcutter', 'carpenter' and 'mason', all integral yet declining workforces.

As the 'truly wild' bees in the British Isles, solitary bees and bumblebees face many of the same threats, but most bumblebees have a better social life. Living in nests of up to 400 individuals, bumblebees adopt a similar division of labour as honeybees. Typical of 'eusocial' animals across nature, this describes a cooperative group in which just one or a few females (the queen or queens) and several males are reproductive. The remaining sterile workers – often numbering in their thousands – provide for the group. Such extreme altruism has long puzzled scientists. From naked mole rats (please Google them immediately) to termites and ants, the militant, organised fortress of eusociality has turned out to be a bit of an evolutionary success story. But only for a very exclusive group of species, which is why we like them.

Some 275 species of bumblebee exist worldwide, 25 of which reside in the UK. We used to have 28, but three became extinct between the 1940s–90s. Such a shame, as they had excellent names: Cullum's bumblebee, the short-haired bumblebee (now being reintroduced in Kent) and the apple bumblebee. Given that bumblebees are properly wild, it fits that most UK species have been riding a tide of decline since the 1940s, under pressure to adapt alongside the changes in farming, urbanisation and associated dissolution of suitable habitat – you know it. Eight species are officially 'endangered', with national surveys observing substantial declines in bilberry bumblebees since the 1990s. You would be right to comment that all insects suffer this same basic hardship as the land has been scoured. British skies, hedgerows, verges, and gardens face one of the worst famines of insects in history. As I'm writing in early June, I think I've seen fewer than five bumblebees in my garden this spring. From the marsh fritillary to dung beetles, it's always the specialised species, with their particular quirks and meticulous associations, that are at most risk. Highly unfair.

I had initially planned to search for the bilberry bumblebee in Northumberland, where it is a bit of an insect celebrity and far better-known than in Devon. Plus, I had always fancied a stroll along an empty, north-eastern beach before retiring to camp in the grounds of a desolate castle, drunk on sightings of puffins – and bumblebees, obvs. But something about the novelty of having a few on my doorstep seemed more interesting. I was ready for it. So, here we go, our last dance.

A thin film of frost clung to the cars. Dense mists robbed the morning air, which smelled different from the previous week. Newer, sweeter. Taking my sweet time to check the tyre pressures on my bike, I was procrastinating to give the sun a chance to burn through, waiting for its hand to creep down the garden wall. Devon's riddle of lanes leading to the moors are sketchy on a cold morning. Tree tunnels are often so dense in canopy by this point that parts of the road barely see sunlight during the day. May 2021 certainly felt confused.

Perching on the gutter above with a piece of tissue in her bill, a female house sparrow cocked her head with such precision her twitches looked almost robotic. Another flew past, so close that I could feel the purr of its wingbeat brush my cheeks. Despite its momentary aloofness, the gala of spring was almost upon us. Food orders triple-checked. Guestlist complete. Anticipation swirled in the mist.

After the first hill, I was already planning stop-off points to have a second (likely third) breakfast. I was riding fast, energised by the morning and being on home turf. Those rich, green tree tunnels offered a rapid-fire of sun-shade-sun-shade, hitting my body in strokes of warm-cool-warm-cool air. There was a lot to absorb.

What's bizarre is that bumblebees are warm-blooded, just like us. Thermoregulation like this is a string to their tiny bows. Strange as most insects are 'ectotherms', so their body temperature equalises with the ambient temperature. When butterflies bask in the sun, their bodies are likely in need of a bit of a thermo-boost, and they instinctively rely on external cues. But bumblebees have a superpower – they can *create* heat and thus be strong, independent citizens of the world's uplands. Presumably, as with most bees, they are also rather busy, so I doubt they have time to sit and sunbathe. And they aren't the exception: dragonflies, hawk moths and a few other big insects can also generate their own heat – they are 'endothermic'. Tiny, furiously busy cogs of flight muscles fire up to produce kinetic energy, the warmth from which is trapped within their oversized downy jacket (in the bilberry bumblebee's case, especially). Generously, evolution has given them one less thing to worry about. Thermoregulation is vital to enable this bee to live the high life at low temperatures. But you may have already joined the dots: who in their right mind would want to be swaddled in fur in a warming world?

Tedburn St Mary, Cheriton Bishop, Dunsford. As England's third-largest county by area, Devon sure puts cyclists through their paces. 'You earn the views here,' someone told me once. On this ascent to the high moors, the hedges were just about low enough for me to see the swell of fields, meadows and woodland, rising and falling around the Teign Gorge. *Where are you now, then?* I kept thinking, wondering what the salmon were doing on their home river. How had winter treated them? Huge, bare swathes of moorland rested above distant hills and freewheeling on my bike blurred it all. I will never tire of these views.

Wooded, almost luminous fields gently swallowed villages, smallholdings and lone houses. A red pillar box and matching phone box saluted a stone cottage in the village of Doccombe like some Great British fever dream. The phone box had been converted into a library of sorts, stacked with old books. A small, handwritten sign encouraged borrowers. Bluebells wobbled with dew. A chiffchaff offered a rhythm for me to pedal against, and somewhere high in the canopy of leaves, a raven croaked gutturally, as though opening an enormous door no one had cracked for centuries. *How much 'England' can you cram in?* I mused. Trying to remember exactly what bilberry looked like, I quickly scanned hedges as if some might be there. *As if,* mate. In any case, going for a poke around the bilberry bush in search of a rare bee sounded like an excellent way to spend a Sunday afternoon.

All this loveliness blinded me to the fact I was cycling in the middle of the lane with a 40-tonne milk tanker barrelling down my rear end. With a noise that sounded like I had inhaled a rock, I rammed my bike into the hedge and waved frantically at the driver in fierce, patriotic apology, bracing the entire side of my leg against a ripe nettle clump.

My teenage catharsis on Dartmoor brought with it the realisation that there are some places you can explore all day without seeing another soul. In the crowding South West, this feeling is becoming rare. Such expanse and Dartmoor's tendency to spread thick fog upon walkers within seconds undoubtedly urged early minds to sprint. The legend of the 'Hairy Hands' of Dartmoor has gripped many. On a dark and misty night in 1921, a worker from Dartmoor Prison lost control of his motorbike on the bridge between Two Bridges and Postbridge (Dartmoor enjoys a good bridge). Since then, numerous and weirdly similar accounts have arisen, telling of large, hairy, calloused hands seizing steering wheels along this same stretch of road.

Arthur Conan Doyle, of course, saw Dartmoor as the natural setting for *The Hound of the Baskervilles*, writing much of the text from Princetown's Duchy Hotel, now High Moorland Visitor Centre. Tales of pixie-clad wells, ghosts, witches and unexplained disappearances lend themselves well to its mystique. Acidic soils create remarkable preservation conditions, putting Dartmoor forward as a leading candidate for western Europe's most important (and complete) area of Bronze Age archaeology. More than 130 medieval settlements offer vital clues to how these communities lived. Stallmoor's stone row monument remains the longest in the world at 3.4 kilometres. And if all that doesn't thrill, then know that its ancient woodlands also house the world's largest land slug, stretching to more than 20 centimetres. (No need for me to be lovesick for Orkney after all.)

More than 160 granite tors scatter the national park, and it's easy to forget how these lumps of rock arrived in the first place. For Dartmoor, granite is as rudimental to its genesis as oxygen is to ours, making it the largest area of exposed granite in southern Britain. When learning about plate tectonics in geography, I heard it better when Dartmoor was the case study instead of Mount Etna. However, my tor diagram bore a closer resemblance to a pile of dung (a very organic one, though, thanks) than an ancient granite outcrop.

Some 280 million years ago, a major mineralisation party was happening beneath the Earth's surface. Like any good party, it was hot, sweaty and sparkly. Molten rock reaching up to 1,000°C cooled and solidified. Quartz, feldspar and biotite minerals crystallised, meshing into a hypnosis of granite. Years of cooling scarred the rock with deep, jagged fractures. More time still and a gradual lessening of pressure from overlying rock allowed the granite to expand upwards, fissuring into iconic horizontal joints. Over 30 million years, Britain's climate transformed from being subtropical and equatorial into an Ice Age. Humans emerged as the dominant land mammal. Dense trees and vegetation relented, then granite formed. Water froze and thawed within its joints. Weather forced its way in, shattering and sculpting the rock into the beloved outcrops we see today.

Rest assured, Dartmoor's millennia of plate jiving has enabled it to uphold a talent for never being the same. With a chorus of tors, ancient woodlands, rivers, peat bogs and acidic grasslands, Dartmoor constantly reinvents itself within a fluid identity. Almost operating under its own microclimate, much of Dartmoor is cooler than the surrounding areas. It could be sunny in Exeter and snowing on Dartmoor – or vice versa. Like Channel 4 in the early hours, you never quite know what you're going to get.

A significant intrigue with bees, of course, is their infatuation with the flower. When it comes to any affair, basic information doesn't suffice: we want all the *detail*, the delicious *subtext* and a thorough who's who. And without bees' extraordinary partnership with flowers, I wouldn't be writing about bilberries, and you wouldn't be reading about them. A world without pollinators is no world at all. Not one that would be fun to survive on, at least. While all pollination is vital, there are two main types: animal (especially insect) pollination and wind pollination. Both methods are crucial for wildflower and crop reproduction and for seeds to grow again and again. Genetic diversity, which we know is indispensable if a species wants to stick around, is upheld by visitations from species like the bilberry bumblebee. Insect pollination is especially well-known for improving fruits' yield, flavour, and reliability, which helps the whole cycle continue and allows warm, buttery beans (other plants are available) to appear on our plates.

It took me a while to get my head around the fact that plants have male and female sex cells (gametes), which they use to reproduce. To the casual observer, a flower's pollen may just be some dusty yellow stuff that stains your nice white trainers, but within it lies a panacea. Male gametes inhabit these yellow grains. When you get pollen on your nice white trainers, you're essentially facilitating floral sex. But bees do it better. When buzzing around a flower to gorge on nectar, pollen grains from the stamen physically stick to the bee's entire body, face and legs. For bees as furry as the bilberry bumblebee, they can leave a flower looking like they've donned a pair of sunny pantaloons, a boob tube and a bucket hat – festival ready.

When visiting the next flower searching for more nectar, the pollen grains from the first are transferred to the stigma, stimulating the growth of a pollen tube that journeys down to the floral ovary, where the female gametes are. As with all fertilisation, this act results in an embryo. It grows into a seed and eventually is contained within a fruit. Seeds must be dispersed via wind, animals, humans or of their own accord so that the entire cycle can repeat. If pollen were stardust, then pollinators would be artists, painting their celestial existence right here for us on Earth. So. Very. Stylish.

This bond between bee and flower is so tight that both organisms have developed and refined complementary techniques to help each other accomplish their missions. Rich scents, colours, designs and nectar are the ingredients of floral seduction. A bee will notice when a flower has made a special effort to dress up and will reward it (in an indirect way) with sex – not a hard one to compute. One of the earliest flowering plants, magnolias likely triggered the birth of insect pollination millions of years ago. Since then, evolution has fine-tuned the strings that pull it together, resulting in the most astonishing harmony. The underlying melody is composed of carefully crafted messages. Strategically placed memos inform a pollinator of vital facts on a need-to-know basis. Guys, a healthy wildflower meadow has more flashing neon signs than Times Square. It's not hard to see why conservationists lament the colossal fragmentation of British wildflower meadows over the past century. All these stories, and their correspondents, lost in translation.

Working together, bees and flowers share an understanding. Forget-me-nots, for instance, have a fleshy yellow ring around the centre of the flower. Acting as a beacon, this draws a bee to the nectar. And get this – once the flower has been pollinated, the plant can feel secure in its succession, triggering this yellow ring to fade to brown. Think of it as a Post-it™ note stuck to the flower reading, 'Hiya. Don't bother. I've been done' to any prospective bees.

Similarly, bees text each other via chemical 'footprints'. Many experiments on bumblebees have observed that once a flower has been pollinated and emptied of nectar, the bee's footprints act as a molecular deterrent, repelling both itself and others. Foraging in this way is more efficient, avoiding flowers with depleted rewards. I would love for this to become standard practice back in our kitchens. Detailed little notes updating us on the fridge's contents would save a lot of time, wouldn't they?

Perhaps my favourite adaptation of all is the secret bee-flower hotline. I am obsessed. In 2013, Professor Daniel Robert led an extraordinary piece of research while I was at uni in Bristol, which discovered for the first time that flowers could communicate to bees via electrical signals. Emitting weak, negatively charged electrical fields towards a bee's weak positive charge establishes a line of electrical communication lasting several minutes. What they talk about in those few minutes is surely the ultimate cipher. How many other Enigma codes are waiting to be broken?

Back to the upland jamboree. Moretonhampstead offered respite (i.e. much-needed caffeine and healing baked goods) before the ascent to the high moor, which began shortly after the Miniature Pony Centre (a day out that would go precisely as you'd expect). Knowing that open moorland awaits makes climbing a long, steep hill more bearable. It is the ultimate reward in many ways because it's a landscape utterly different from the one you've just passed through – this is typical of the British Isles, to be fair, which hold worlds within worlds. The vast, bare moors that had rested above earlier hills were now nearly a foreground. Bike and I were gifted heady views across west Devon at every farm gate or gap in the hedge.

Swerving to avoid something long and dark in the middle of the road, I realised it was an oil beetle. A tiny head and thorax rested on a tremendous abdomen, segmented like window seats on a train carriage – it looked black from some angles, violet iridescent from others. Crossing footpaths to a new foraging patch isn't unusual for them, but this lane must have felt as expansive as a midnight M25. I sped on.

With the landscape continuing to unfold like sections of an old map, other cyclists and walkers arrived on the scene. Cuckoos, too – our visitant charlatans. Their unmistakable call and response is all we want from summer, isn't it? Amazingly, the cuckoos that spend summer on Dartmoor have flown 5,000 miles from the Congolese tropical rainforest, to which they will then return. The road was opening up. Ewes ushered their lambs idly across. Two ruffian ponies figured the best grass was the stuff overhanging the road, and their random wanderings caused cars to pause, drivers watching curiously, smartphones aloft. Sooner than expected, I was on the moor. The sudden absence of the Devon hedgerow meant I digested immense barren vistas on the move. Up ahead sat The Warren House Inn, well known for maintaining smouldering embers burning in its fireplace since 1845. It also happens to be where, rather endearingly, my brother inadvertently let slip the name of my first nephew a couple of months before he was born.

I had a very specific destination this time – so specific that I had been given a grid reference for it. Serious bilberry bumblebee hunting was afoot. Mike Edwards knows about bees to the same depth that one of my cousins knows about Ed Sheeran. Regularly cited as one of the UK's top bee experts, he coordinated the Biodiversity Action Plan for bees across the UK and co-authored two of the most prominent field guides. It was Mike who had recommended I visit the Dartmoor bees rather than the Northumberland lot. A top tip, as it turned out – thanks, Mike.

'We found many queens coming to the flowers,' he had told me over the phone before I set off. I tried hard to visualise it. South-west of Birch Tor and just off of the B3212 lie the remains of some old stone workings. Between the late eighteenth and nineteenth centuries, this whole area of North Bovey was a tin-mining powerhouse, occupying the largest mines on Dartmoor. Invisible from the road, shafts and earthworks remain within the furrowed brow of the hill in this peculiar confluence of the natural and the artificial where our bilberry bumblebee was (hopefully) awake following hibernation and feeding on something pretty.

'What are you trying to do exactly?' Mike was animated and inquisitive. He not only understood bees very well, but people too, recognising our particular (apologetic?) way of looking at things if we dare to care. During the early 2000s, Mike led various surveys for the Bumblebee Working Group (BWG), a handful of which he passed on – 'just if you fancy a bit of light reading …' I suppose I did. Being a grouse moorland in a previous life, Birch Tor's current medley of flowering bilberry, heather, white clover, birds-foot trefoil, grasses and bracken evidences its recovery. But between 2003–05, surveys of bumblebees around Birch Tor found that the bilberry accounted for just 3.4 per cent of other bumblebees surveyed. Such low numbers prompted further observation of Dartmoor's share. Pockets of nectar-rich sallow[37] nudged Mike and others to keep looking for queens around Birch Tor after they were seen feeding on sallow in Abernethy, Scotland. Their persistence was soon rewarded. 'We soon saw *B. monticola* queens working the catkins, and with considerable gymnastic efforts, we were able to capture a number of queens and remove their pollen loads for later analysis,' the report reads. Such analysis is

[37] A species of native willow, also known as 'pussy willow'. Behave.

critical if we are to understand what flowers the bees are feeding on – especially in the dawn of spring.

Considered short compared to most bumblebees, the bilberry bumblebee life cycle lasts around four months. Triggered by rising temperatures, queen bees emerge from hibernation in late April, following a winter underground. Understandably, the post-winter mood is delicate. They must replenish their energy ASAP, and before they can even start thinking about sourcing a new nest for this year's brood. Polylectic foragers, bilberry bumblebees can collect pollen from a wide variety of flowers, the protein from which is vital for larval development. Handy. But fieldwork like Mike's has shown them to consistently favour bilberry, and have a small number of specific favourite plants throughout the year. There is method in the upland medley. Only those plants with the richest nectar stores have evolved to flower to *coincide* with the emergence of the queen. And, like any good marriage, the bilberry bumblebee vows to support their growth, till death do they part.

When was the last time you thought of a bee's tongue? You and me both. As far as tongues go, bilberry bumblebees have short ones, limiting them to shallow-tubed flowers that don't require a cave-diving exercise. Mike explained that 'what they need is open moorland with successions of flowering from the end of March onwards'. He added that, in good habitat, you could predict the sequence. First, they aggregate on sallow, which flowers early. An early spring bloomer, the tiny yellow bombs on the male sallow flowers promise huge nectar #gains for emerging pollinators. And fun fact: sallow nectar was once ranked second in the top 10 plants for nectar production in a survey for the UK insects Pollinators Initiative. Ooh err.

Next 'the bee will move (somewhat predictably!) to bilberry, birds-foot-trefoil, white clover, raspberry, bell heather, sometimes bramble'. Mike explained how the bilberry bumblebee's habitat is a 'landscape-scale thing' and how essential this continuity of food stations is for their upland survival. I'm sure we can empathise? The majority of us are fortunate enough to receive meals at regular intervals throughout the day, the week, month and year. Any disruption to this, however, is hard to stomach, as it were. But if this daisy chain of events severs for this bee, they can't just mend it by shopping elsewhere.

'They are part of the old upland system. We can cry over it as much as we like, but it's gone,' Mike stated. 'You're left with isolated pockets of good habitat and not much in between.' Remember, this is our mountain bee. A bee synonymous with: 'elevation', 'peak', 'ridge' and 'range'. A bee craving open space to forage. 'Yet it won't thrive in pure moorland alone,' he reminded me, 'it needs rich grass – *other* habitat – as well as bilberry heath.' I can appreciate that. Unfortunately, sheep are very partial to a romp about the bilberry heath, too. How very predictable.

Farmers have been invaluable guardians of Dartmoor National Park for more than 6,000 years and continue to graze native livestock across 90 per cent of its land. The whiteface and greyface sheep have featured heavily from Dartmoor's Iron Age to the present day. Around 150,000 of them graze these uplands annually, bred for their meat and wool, but changes are coming. As I write, a substantial rewilding campaign is competing with tradition. In autumn 2020, Natural England ordered the removal of 1,000 sheep from the moors around Okehampton to allow upland flora, including bilberry, to regrow. A huge debate ensued. Although removing sheep doesn't remove the problem of reduced upland diversity, it's worth considering when we're looking at the bilberry bumblebee.

Because bumblebees only tend to store small amounts of food at the colony, they need to have flowers available to fund regular foraging trips – every few days or so. Some of us prefer braving Tesco once a fortnight to smash out a mammoth shop; others get what they need every couple of days. Along with my dad, bumblebees are more frequent shoppers, and concerns are that continued overgrazing could rapidly escalate an 'ecological separation' of this bee from its Devonian frontier. Food shops become more erratic. Frays appear in that damn daisy chain.

Some bumblebees can access flowers in 'illegal' ways, biting a hole and robbing nectar from the back of the flower. Although relatively common, scientists think that this 'nectar robbing' increases during shortages of suitable flowers upon which to feed. Research from the University of Sussex has observed how these sneaky tactics can transmit through a population and become learned behaviour – potentially very useful! But all of this adaptation takes time, which we don't have. Although I admire the bees' ovaries (and brains!), they shouldn't have to stoop to burglary to survive.

The 2000–2005 surveys I read by the BWG persuasively encourage the need for mass flowering and soft, rotational grazing – directly linking upland grazing pressure to the absence of the bumblebee. And yet here we are, nearly 20 years on from these first recommendations, having the same conversation. Mike stated simply how winter sheep grazing removes the tips from the bilberry and reduces alternative flowers. Rookie mistake. Yes, large areas of Dartmoor do have bilberry, 'but no other flowers – so the bee just doesn't occur in those areas'.

Helmet, rucksack, bike and self were strewn across a bank by a small moorland car park, as though the four of us had bailed from a helicopter into some disaster zone. An all-day breakfast-in-a-bag simmered on a camping stove alongside some Mexican rice for lunch. You should know by now that most of my meals blur into one. I was a day-tripper in a good place. Something black flew across my face, buzzing loudly before landing near my outstretched leg. I prayed for a red bum, but it was as white as a mountain hare – buff-tailed. *Bombus terrestris,* the bilberry's bigger, omnipresent cousin, finds success in an impressive range of habitats across the UK and Europe. As I spooned my lunch hurriedly, I couldn't stop hearing, 'Bzz....ZZZ...zzz..*zzzzzZZZZ*'. Whether it was, in fact, more bumblebees, my mind or the motorbikes that raced along the B3212, I can't be sure.

I began my beeline (wahey!) to the epicentre of Mike's grid reference. As with all the species in this book, the odds of *not* seeing this bee only made me want to see it more – I fall pathetically for the ones that play hard to get. Somewhere nearby, an outpost of this bumblebee has persisted, despite unimaginable challenges. Pride craved confirmation. I wondered if the area around Birch Tor was their most southerly range in the UK, meaning that it was genuinely isolated, and therefore, screwed?

I threw this thought at another professional bumblebee person – possibly the most go-to bee guy in the UK, if not Europe. The last time I spoke with Professor Dave Goulson was in a meeting where we mused the possibilities of whether beavers could help create a habitat for bees. This time it was *all* about the bees. Animals which, as you may have gathered, are so mind-bending that they deserve a book of their own. Aside from being a leading academic, speaker and the author of several brilliant books, Dave also co-founded the Bumblebee Conservation Trust (BBCT) in 2006. Responding to the crash in UK bumblebee populations, he and Dr Ben Darvill have built a science-led organisation. Over the years, the BBCT has gradually filled in vital data gaps on Devon's bumblebees as part of the West Country Buzz Project. Before this, my county's bumblebees were among its most under-recorded species. Dave and his team were also central to improving understanding of those foraging footprints that bumblebees leave behind. Safe to say, Dave is a legend.

'So, bumblebees have declined to a point where you would never find many species unless you go specifically to one of the tiny surviving populations,' he told me. 'That's what got me into it all, I guess – that there were many bumblebees in the books that I just couldn't find. I became hooked on the idea.' Dave revealed that the bilberry bumblebee was fast becoming one of those species, though not necessarily because of direct habitat fragmentation, which is the nightmare for most bumblebees operating around meadows. 'No, it's more that the bilberry is associated with habitat that's not easily farmed,' he observed. 'Moorlands haven't changed as much as the surrounding countryside. They're still being fucked up, obviously, but on the whole, they've been spared the devastation of the lowlands.'

It appears that our mountain bee has other hills to climb (ones they cannot see). Their fragility here on Dartmoor is the writing on the wall. Sizable (but colourful) capital letters reading: ATTENTION PLEASE. We know this upland storyline – we could hold hands and recite it around the dinner table. To some extent, mountain hares, merlins and marsh fritillaries are all singing from the same tatty song sheet, written by the same tired lyricist. As with all endangered species, bumblebee distribution surveys are vital. If we know where they are, we can try to figure out how to help. The coastguard cannot rescue someone dangling from a cliff if they don't know from which crag they're hanging.

During our call, Dave added: 'You'd expect distributions to shift towards the Arctic – that's what species did for millions of years during historical climate change. But what is worrying is that the southern edge of their range is shifting, but the northern edge is stuck.' Dave cited an extensive study in 2015, which revealed a disturbing stagnation of bumblebees worldwide, caught in what researchers call a 'climate vice'. Further research in 2018 predicted these climate change-related losses among bumblebees are on standby to escalate. Focusing on North American populations, the researchers observed a trend in significant range contraction, even when bees could disperse. Dispersal is simply not fast enough to match the pace of climate change already underway. Bumblebees *should* be able to move north to keep cool, but they're not.

Talking to Dave, I was reminded of something Mike had said. 'People often think that these things can't relocate – but yes, they damn well can. The bilberry bumblebee is the most amazing wanderer! Colonising Ireland from south-west Scotland about 30 years ago. In fact, if these bees can't move, they're dead.'

So, if they can relocate, what's stopping them from packing up and summoning Pickfords? 'My guess is current populations are just too low, and the habitat isn't there to support them.' Dave nostalgically referenced the past when wild insects would have existed in enormous, fluid populations. 'If they needed to move geographically to cooler areas, they could do so – thanks to the continuity of habitat. But, now? It's getting claustrophobic.' In this grim gamble, if the density of suitable habitat thins, the chances of reaching it thins too. 'They've somehow got to hop over a sea of crappy habitat simply to reach the next best site. If that exists,' he added darkly.

Absentmindedly, my hands trailed pinches of flaxen grass on the rough track leading into the heart of the stone works. I was thinking of my sister-in-law, Beth, who, at the time, was heavily pregnant with my third nephew. When we met, my first thought was that her hair was the most beautiful I had ever seen – poker straight and shades of sunlight, strikingly similar to the grasses gliding through my fingers. Struggling to focus, I turned right down a new trail and fiddled with my nails, unsure why I was feeling uneasy.

Clouds assembled. The temperature dropped considerably around the edges of the mine. *Please, don't,* I pleaded, knowing full well that pollinators would retreat with rain. Had this spring been more reliable, bilberry (or 'whortleberry' in Devon-speak) would be out by now. But I had a hunch that I was bumblebee hunting too early. Yet deep in the moorland understorey and surrounded by pussy willow (shush), the air felt balmier, as if we were in a wetland. As a subterranean species, queen bilberry bumblebees often recycle old nests of field voles and mice, which occupy a similar habitat. I could imagine a vole enjoying this place. *Must be close*, I thought, jaw clenched, checking the map. Ribbons of stream wound undefined routes around the track. I was alone. Thousands of male sallow catkins were held in a halo of pollen, which, upon further investigation, were compacted into hundreds of little balls, like the tips of a hairbrush.

Whipping my head around like some agitated dog, I was foolishly trying to pinpoint the aggressive bee flying in and out of my head. Impossible. Tired legs tickled with phantom itches, and my patience was ebbing. Distracted and irritated, I noticed a baked bean was drying in a strand of my hair. I circled clumsily behind some rocks like a wheelie bin, eyes fixed skyward, a low coil of blackthorn shredded my ankles. Frustrated, I wiped away the blood roughly with my hand.

But then I noticed them – a thousand phones vibrating. Looking up, I realised I could hear the *entire* tree humming – a deep, thrumming baritone, supported by tenors and basses dotted throughout their canopy bandstand. One hand on its trunk, the other shielding my eyes, I gasped because a tiny black bullet was darting around the sallow far above me in the canopy. A miniature spaceship entirely consumed by its mission. And then another. And another. Hundreds of bees of various types had been here all this time. Feeding. Pollinating. Living. I just wasn't listening.

This bee is completely orange. I have blurry memories of drinking Aperol spritzes with my lifelong pal, Rachel, that had me wobbling around the streets of Majorca in search of pasta. Bees were commuting fast and high amid the network of branches, making it impossible for me to differentiate between them. A thousand wing beats differing in pitch and cadence made it busier than London Paddington at rush hour. You could *feel* the urgency of spring.

Writers have observed how the bilberry bumblebee doesn't 'bumble' about like other bumblebees. Instead, it works the understorey slowly, methodically, as a carpenter might over their workbench. I had finally locked on to a queen, radiating unmistakable warmth like a deep, rusty sunset. She was a big girl, too. She lobbed off a sallow flower that fell into the stream with an audible *plop*. *Take me to your leader*, I thought, imagining some resplendent queen of queens silently observing this scene from her nearby throne. The bilberry bee I watched seemed to work a patch fully, then move on, flying low like a chinook on a classified assignment. Buff-tailed were far more numerous, and nippier too. A bar of signal brought a text from Tom sent at 1.28 p.m. telling me that Beth had given birth to their third baby boy. I beamed under my bee tree and contemplated what it would feel like for this rare little world to swallow me up, just for a bit.

As 'pollen storers', bilberry bumblebees actively feed the developing larvae back in the nest from a central pollen store – a larder – rather than wrapping each larval cell with a little pollen party bag as solitary bees do. 'Pollen is a complicated food source with a dual purpose: food and protein for larvae and the chemical constituents to fertilise the flower,' according to Dr Beth Nicholls – a research fellow at the University of Sussex and formerly part of Dave's research group. The bumblebee world is quite niche. This Beth runs the Nicholls Lab with researchers specialising in the cognitive mechanisms driving foraging choices in bee populations. Permission to be inspired. Young, breezy and sporting an excellent genre of headband, she chatted with me over Zoom. Thirty seconds in, and I aspired to go for a drink with her. Two minutes in, and I considered asking if she had a job going. 'People used to think that insects had rigid behaviour patterns, but bees are learning about flowers all the time! Even these tiny animals can adjust their behaviours based on the rewards.'

I asked Beth if there is plasticity in the bilberry bumblebee's foraging behaviour. Could it modify its floral needs to suit a shifting upland life? She told me they could do so, but ultimately their morphology limits them – big bumblebees can struggle to visit smaller flowers and are restricted to specific locations. 'Dave taught me how strange bumblebees are. Most insect groups become more species-rich towards the tropics, but bumblebees just aren't like that. If it's cold, they can shut down and almost re-enter hibernation. But if it's warm and wet, you're still burning energy, but you can't get out. That's the primary concern – the colony can collapse quickly.'

There are growing fears about this bee and its hotness. An unusual thing to worry about, I'll admit, albeit well-founded. The good news is that, unlike merlins, for instance, bumblebees (generally speaking) have research on side. Multiple studies have investigated how exactly bumblebees can withstand harsh, cooler environments. As the name suggests, *Bombus polaris* prefers life at the Arctic Circle. *Bombus kashmirensis* enjoys a romp around the Himalayas. Studies also admit apprehension when considering how these upland insects will stay cool as the world warms. 'They're going to be hit harder than most,' Dave had told me. Their dog days are only just beginning.

Earlier, I mentioned how bumblebees generate their own body heat to navigate these extremes. 'Bumblebees can only fly when their internal temperature is between thirty to thirty-five degrees Celsius,' Dave said casually. I was astonished. 'Yeah! You could watch a queen flying around in air temperature that's near-freezing, but if you were to measure her internal temperature, she'd be thirty-plus degrees.' I've been dreaming of miniature thermometers ever since.

As part of their pre-flight checks, all bumblebees must warm their flight muscles – 'pre-flight thermogenesis', if you'd prefer a fancy term. Ironically, this means they shiver. Beating their wings around 200 times per second generates heat. Around 40 per cent of the average flight muscle volume is crammed with mitochondria – the organelles that nearly all living organisms have to generate energy in respiration – our tiny batteries. Shivering essentially activates the bee's mitochondria, firing up the muscles and initiating lift-off.

Sometimes, the bees can temporarily overheat – and that's OK. 'They can just pump hot haemolymph [insect's version of blood] from the thorax back into the abdomen, cooling them down,' Dave explained. But there's a limit. 'Once they start flying, they cannot stop generating heat whether they want to or not,' he continued, 'so if the air temperature is high, they overheat. They can't fly; they can't collect food. They waste time at the nest waiting for conditions to improve, which means they aren't pollinating.'

This situation could be serious for two reasons. Back in 2003, Dave found evidence to suggest that bumblebees are more effective pollinators than honeybees, as their ability to withstand the cold gifts them more hours in the working day – and year. Dawn, dusk and cool summer spells are all on the cards. As a group, wild bees are twice as effective as honeybees in pollinating crucial crops, like coffee, almonds, strawberries and oilseed rape. On the flip side, Dave has observed how a reduction in pollination for any bee can potentially trigger an 'extinction vortex'. Flower reproduction suddenly isn't as rampant. Other pollinators go out of business. In worse-case scenarios (yet scarily simple economics), both bee and flower populations can crash. Poof! It's over. Nothing to see here.

The trouble with the bilberry bumblebee is that although it's a 'gobsmacking species to look at' (Mike's words), its furry coat is its Achilles heel. Being one of the hairiest bees, the maximum temperature that it can fly at is likely lower than that of lowland species. 'Warm temperatures can increase larval development back at the nest up to a point, but too hot? Like recent summers? Lethal.' Mike paused. Cripes. Thinking back to all the times I have been too hot, and I remember how much I hate it: racing through London to catch my train to Scotland; again on that train to Scotland; that time I made the fatal error to wear jeans and a jumper in Wilko when the heating was on full blast and the escalator had broken – irritable and lethargic in all scenarios. Overheating is universally rubbish and, at times, deadly for us too.

As I write, in July 2021, British Columbia, Canada, is currently in a roasting oven of a heatwave. In what scientists are referring to as a 'heat dome', record temperatures of nearly 50°C are not only causing sudden human deaths but that of almost 1 billion marine animals in the waters off the Canadian coast. More than just a global warning, it's a real-time demonstration of the vast amount of heat that is increasingly being trapped within Earth's atmosphere. A report from NASA and the National Oceanic and Atmospheric Administration (NOAA) presents a sobering picture of a stark imbalance of solar radiation emerging between 2005–19. During those 14 years, solar energy gain on Earth has nearly doubled. It would be like turning an oven preheated to 180 °C up to 280°C just as your spuds are almost cooked. Would you prefer edible roasties or a charred mess?

Back to the bilberry bumblebee. Fabulous fur coat or not, I imagine the anguish of being unable to remove it when temperatures are rising would be similar to wearing orthodontic braces forever. When you first have them fitted, they hurt like hell. You're so desperate to rip them off that you dream of it. But, no can do, hun – you're trapped in a vice. At least the climate vice we're stuck in is temporary. It is, right?

Bumblebees play a critical role in our environment. They're an ecological linchpin. Don't even bother questioning that. But I've deliberately not told you their monetary value. I don't see the point. Yet later, we start asking other questions when we decide that a bee as, a bee, is no longer enough. *So, how much are they worth? Yeah, but how much money do they bring in? If they don't contribute X, then what's the point?* Sadly, this angle has become a crutch when trying to communicate the natural world within the confines of a capitalist one, because mixing everything in the cocktail of the economy is easier. I did it in the dung beetle chapter, hoping that linking their services to money might prick the right ears. Open a bottle, grab some hors d'oeuvres, put a number on it, dammit – *then* Whitehall will listen!

Naturally, many articles on pollinators begin by quoting the fiscal value of their services.

Even over the screen, I saw Dave wince. 'What do bees *do* for us? That utilitarian argument has not aged well.' He shook his head. 'I absolutely hate that approach. It's shallow and actually just really sad.' He told me there will always be people for whom the economic argument of bees is the most powerful. There is a widespread attitude that if insects don't have a demonstrable value to humans, it is beyond our remit to care. 'But I'd love to do more to challenge this and stop justifying conservation on the grounds of how it benefits us. Dodos didn't do a fat lot for us, and our society hasn't collapsed without them. But it's a poorer world we live in without them.'

Beth Nicholls, too, shared a similar view. 'Asking what 'the point' of an animal is – especially insects – is such a strange attitude.' We were both fiddling with the pendants on our necklaces. Deep in thought. 'Like, what's the point of us then?! We cannot be so single-minded.'

So, what if the bilberry bumblebee was to become extinct? Total disaster? Or just a shame? According to Mike Edwards, 'As long as something shoves its face into the bilberry plant and pollinates, we're not going to be in an environmental crisis if this bee were to die out. The buff-tailed would probably fill the gap quite well.' In the same vein, Dave told me it 'wouldn't make a blind bit of difference'. Bilberry bumblebees, specifically, look lovely. But there's no evidence that they pollinate any of our crops. So, to be crude – what *is* the point?

Mike considered this for a moment, then threw it back to me. 'Do we want to see the same thing every time we go somewhere? People go and watch the same football clubs forever, but do they want to watch the same games forever?' Hello, moral compass? Be a babe, and tell us what to think? An upland without the bilberry bumblebee lighting up the understorey would be a boring one indeed. Not worth the money, that's for sure.

My head hurts. I'm worried that we risk making a mockery of bumblebees. We are blithely overusing their looks as though they are ornaments of the countryside rather than a cornerstone of it. Bumblebees have become our go-to species when it comes to greenwashing our glorious* (*broken) *system*. However, the future of the bilberry bumblebee and its Aperol arse can be bright, but only if we get off our own.

It appears that we already know what to do. 'But we need to fucking wake up and stop dithering about,' Dave stressed. He was referring to the National Pollinator Strategy introduced in 2014 and how little humans have achieved since, bar a national monitoring scheme. 'Counting pollinators is important, but that's not going to save them. Paperwork isn't going to walk out of Parliament and save bees. No, we need to galvanise rapid change, and we need support to do that.'

We need to stop diluting and start concentrating. Take risks. Taylor Swift knew what she was doing when she told us to be *Fearless*. Just like all the species I've introduced throughout this book, relentlessly taking risks to survive. There is a strong case that the more diverse a pollinator community, the greater its resilience. The more varied bacteria that children are exposed to, the more robust their immune system becomes. Simply releasing captive-bred bees back into a landscape to make the data look better is counterproductive, as M&S accidentally proved early in 2021. Although well-meaning, their plans to release 30 million honeybees into the environment was not an answer. Aside from the high possibility of out-competing wild species like the bilberry bumblebee, honeybees can also introduce deadly invasive diseases. Let's just say wild bees (and their researchers) are not species you want to piss off.

Of course, we won't implode if the bilberry bumblebee disappears tomorrow. But there's no denying that it colours the upland tablecloth, probably to an extent that we will never fully understand. Broad-scale habitat management across hotspots like Dartmoor can help the bilberry bumblebee to keep up.

As I write, conservation organisations are pursuing a renewed zest for collaboration following the outbreak of the Covid-19 pandemic. We're realising that economic and societal restoration is incomplete without restoration of the environment. Across the Peak District National Park in 2019, the BBCT, the National Trust and the RSPB planted 1,000 bilberry plants inside sheep-proof cages to protect them from being grazed. Ecologists think 10 megatonnes of carbon reside in Dartmoor's soils alone, equivalent to the annual output from all UK industries. Experts are urging land managers to seal cracks in these moors and prevent carbon leaks. They are also highlighting all national parks for their crucial role in accelerating widespread healing. It's a shame this realisation has arrived at such a tipping point, but it's better late than never.

'We must think of things that everyone can do,' Beth reminded me. 'You can do a lot simply by providing food for bees and planning your planting, so there are flowers throughout the year. And please – stop pulling up dandelions!' She laughed.

The shocking revelation is that *we* can build that damn daisy chain. Yes! Each one of us. We can shimmy those baselines back. Plant mini meadows on balconies, window boxes, pots and grow bags. Remove 'weeds' from the vernacular. Be a little messier in a tidy world. 'And then you can sit outside with a gin and tonic, with bees flying past, knowing that *you* helped them get there,' Beth said triumphantly. Sounds good, right? Let's not forget that the beauty of the bee is that it's one of the few things that every community, every school, every household – and *you* – can rescue.

'We shouldn't give up. We can't,' Dave said simply.

I thought it would be fitting for us to end the book with a bee. Thinking back, I don't think I even spotted an actual bilberry bush during my little jaunt. I cycled home in a complete trance. So, whether Dartmoor's bumblebees made it through May's cold, wet weeks, I'm unsure. But I like to think they did. After all, evolving as our mountain bee must equip them with some serious grit.

I'm writing these final thoughts cross-legged, wearing one of my dad's old T-shirts. It fits me like a dress. England is simmering in yet another heatwave, and I'm struggling to stay cool at my desk. It's been a few months since I finished my travels. Since then, I've seen a fair few beavers, some jellyfish and rediscovered that I could be a professional dancer (when alone). Whenever a song comes on by The 1975, there I am – straight back on North Ronaldsay – feeling poetic. Been out on the bike a bit, though not as much as I would've liked.

If I'm honest, I'm feeling a little lost. Does this final chapter end my time with these 10 species, their landscapes and the people fighting for them? Our distracted lives make it all too easy for us to look for what's next instead of seeing what's here. I'm worried that in moving on, I'll forget them, but their story matters.

Where to now?

EPILOGUE

The Actual End?

They tell me an epilogue is a good place to answer questions. Perhaps I should use this opportunity to tie up loose ends and make an effort to leave you, dear reader, feeling satisfied. I should inform you of the fate of our main characters – the 10 species – perhaps telling you where they are now and muse where they'll be in a decade. Well, the truth is – I have no idea.

I could have fun crafting a sweet nirvana for our planet, where fossil fuels are gone, insects resume in their victorious clouds, the oceans heave with life, and everything has (at last!) returned to equilibrium. Or, I could spell out the extinction of humankind and the death of the planet as we know it. Both futures are possible – we just have to decide which one we would prefer.

These 10 species have taken us on quite the ride, and yet we've barely scratched their surfaces. I think I might prefer it if, after reading this book, you are left feeling *dissatisfied*. I would prefer that you feel fidgety – a little restless. Maybe, like me, you're seeking closure and wanting more answers than ever? As you've seen, we've got incredible, smart and passionate people seeking those answers as well. Over the next decade, we're going to arrive at more crossroads. We're going to have to make tough calls, and we'll need to lean on science like never before. We're going to have to help nature help us.

Taking inspiration from our 10 lead characters, we're going to have to rise to these challenges and adapt accordingly. Each of these species has been doing this for millions of years, so we'd be wise to pay attention. The good news is that, in most cases, we know exactly what to do. We should have more faith in ourselves! So, in my hunger for us to be braver, I'd like to leave you with two questions:

What do we want from the world?

And what does the world want to be?

Acknowledgements

It's very strange writing a book during a pandemic – especially when it's your first one. For me, *Forget Me Not*'s journey has been made unforgettable thanks to the people along its path (of which, I'm delighted to say, there were *many*). The weird thing is that I've barely met any of them. As you've read, most interaction was remote. And yet, I cannot express how grateful I am for their time, care and unrelenting interest in helping me navigate these complex stories. These 10 species had the power to bring people together during a crazy time, and I think that's quite something. Safe to say, many a wine is owed.

As I outlined in the introduction, this book is my tribute to the science and research that is keeping our planet alive. In no particular order, I would like to especially thank the following people and organisations for their invaluable work and contribution to this book: Jo Poland, Dr Caroline Bulman, Rachel Jones, Holly Dunn, Dr Hanna Nuuttila, Dr Richard Lilley, Evie Furness, Jake Davies, Jack Merrifield, Dr Orly Razgour, Craig Dunton, Dr Holger Goerlitz, Heather and Alison Duncan, Rory Crawford, Dr Elizabeth Masden, Dr Daniel Johnston, Chris Jones, Issy Tree, Charlie Burrell, Darren Mann, Penny Green, Mary-Emma Hermand, Rebecca Varney, Eleanor Lewis, Dr Janina Gray, Denise Ashton, Dr Jamie Stevens, Peter Cairns,

James Shooter, Dr Marketa Zimova, Dr Isla Hodgson, Dr Scott Newey, Jack Baddams, Dr Alex Lees, Mike Price, Indy Greene, Ed Drewitt, Jamie Dunn, Dr Matt Stevens, Professor Dave Goulson, Dr Beth Nicholls, Mike Edwards, Dr Isaac Brito-Marales, Dr Andrew Read, Dr Anna Sànchez Vidal, Dr Margaret Kadiri, Professor Richard Unsworth, Sten Asbirk, George Monbiot, Back from the Brink, Butterfly Conservation, The Wildlife Trusts, Salmon and Trout Conservation, Wild Trout Trust, Rivers Trust, Mammal Society, Sea Trust Wales, the RSPB, the BTO, Hawk Conservancy Trust, Project Seagrass, Knepp Wildland Safaris, North Ronaldsay Bird Observatory, Bumblebee Conservation Trust, Bat Conservation Trust, Trill Farm, Woodland Trust, The National Trust.

A special thanks to all the universities and institutions that have conducted the research on which this book leans so heavily. Particular thanks to Cross Country Trains and Great Western Railway for safely carrying a tired, anxious Sophie around Britain.

Thank you to Dr Ruth Tingay for lifting me away from the dreaded imposter syndrome and consulting so diligently on a couple of chapters. Thank you to Ben Macdonald, who advised me at a crucial, exhausted moment to, 'channel my inner Djokovic'. To Rachel, Lucy, Harri, Briony, Joe, Hannah (Gina), Nina and all my wonderful friends, the people I've worked with and those on social media who I have never met - *thank you all* for your kindness, brilliance and humour.

Thank you to my Beaver Trust family for your unending love and support. To Agent Sabhbh and Curtis Brown for coming on board so gallantly in the middle of the madness. To Abby Cook, who has brought our 10 species to life. And, of course, I have infinite gratitude to the magnificent team at Bloomsbury. Not least to Julie, my fierce, incredible editor, for giving me the freedom to write the way I wanted to (and for teaching me not to write like Yoda). Thank you to Jim for taking a chance and changing my life on that dark January day. I'm still pinching myself.

I genuinely couldn't have written this book without my family. To Tom, Beth and the boys for bringing me so much joy. Also, thanks to my wider family for your support – Donna and Nev, in particular.

Parents – you are both an inspiration. Thank you for being the best friends I could have asked for throughout all of this. Love you so much. This one is for you.

And finally, thank you, Jacob. Since we met on a minibus in a car park 11 years ago, you have stood by my side as we've grown up together. I couldn't be luckier.

Stuff To Look Into (if you fancy it)

I did a rather gross amount of research for this book. Mainly because I barely knew anything about each species when I started writing and also because I wanted to get the best out of the brilliant experts who kindly gave me their time. I felt ridiculously lucky to be able to spend so many months learning. I honestly don't think I've enjoyed anything more.

But I deliberately haven't included a list of references here because I don't think that's why you have bought this book. Instead of proving my Googling skills to you, I'd rather you hang on to the stories and feel it's been worth the read.

Please – be rough with this book, stuff it in your bag, bend back the pages, share it with friends and take it on some trips of your own. These species and their habitats need all the love they can get.

However, if you do have any questions about my sources of information, please ask – I have it all ready to ping over in a link to you and would be happy to chat about it.

Come play on social media (so trendy!): @sophiepavs

What's next?

I don't blame you *at all* if you're feeling that I have somewhat bombarded you with information – some of which is on the heavy side. At times during writing, I've felt disorientated and very 'but what do you expect *me* to *DO* about it?!' I hear you. So, below are a selection of signposts to utterly brilliant material that may feed those thoughts and point you in a helpful direction:

Prepare to be inspired by the work of:
A Focus on Nature: afocusonnature.org
Back from the Brink: naturebftb.co.uk
Bat Conservation Trust: bats.org.uk
Beaver Trust: beavertrust.org
Bumblebee Conservation Trust: bumblebeeconservation.org
Butterfly Conservation: butterfly-conservation.org
Cairngorms Connect: cairngormsconnect.org.uk
Hawk Conservancy Trust: hawk-conservancy.org
Heal Rewilding: healrewilding.org.uk
Knepp Rewilding Project: knepp.co.uk
National Oceanic and Atmospheric Administration: noaa.gov
Mammal Society: mammal.org.uk
Project Seagrass: projectseagrass.org
Rewilding Britain: rewildingbritain.org.uk
Salmon & Trout Conservation: salmon-trout.org
Scotland The Big Picture: healrewilding.org.uk
Sea Trust Wales: seatrust.org.uk
The British Trust for Ornithology: bto.org
The National Trust: nationaltrust.org.uk
The RSPB: rspb.org.uk
The Woodland Trust: woodlandtrust.org.uk
The Wildlife Trusts: wildlifetrusts.org
Wild Trout Trust: wildtrout.org
World Wildlife Fund: worldwildlife.org

You will learn a lot here:
Kate MacRae/Wildlife Kate: wildlifekate.co.uk/my-blog
Kate on Conservation: kateonconservation.com/2019/12/31/top-10-wildlife-bloggers-2020/

Low-carbon Birding (great for individual action ideas): lowcarbonbirding.net

North Ronaldsay Bird Observatory: northronbirdobs.blogspot.com

Raptor Persecution UK (Dr Ruth Tingay's blog): raptorpersecutionscotland.wordpress.com/about

A gentle nudge to read:

Back to Nature by Chris Packham and Megan McCubbin (Two Roads, 2020)

Bringing Back the Beaver by Derek Gow (Chelsea Green, 2020)

Diary of a Young Naturalist by Dara McAnulty (Milkweed Editions, 2020)

Feral by George Monbiot (Allen Lane, 2013)

Forecast: A Diary of the Lost Seasons by Joe Shute (Bloomsbury Wildlife, 2021)

H is for Hawk by Helen MacDonald (Jonathan Cape, 2014)

How to Be a Bad Birdwatcher by Simon Barnes (Short Books, 2004)

Humankind by Rutger Bregman (Bloomsbury, 2020)

Rebirding by Benedict MacDonald (Pelagic, 2020)

Sapiens by Yuval Noah Harari (Harville Secker, 2014)

Save Our Species by Dominic Couzens (HarperCollins, 2021)

Selfie by Will Storr (Picador, 2017)

Silent Earth by Dave Goulson (Jonathan Cape, 2021)

The Overstory by Richard Powers (William Heinemann, 2018)

The Salt Path by Raynor Winn (Michael Joseph, 2018)

Wilding by Isabella Tree (Picador, 2019)

Excellent ear food (podcasts):

BBC Earth

Costing the Earth
Desert Island Discs (if you want to be reminded of humanity's brilliance!)
For What it's Earth
Into the Wild
Outrage and Optimism
Radio 4's NatureBang
The Lodge Cast (shameful plug!)

If you're more of a ⋆free app⋆ person:
Bee Count
Birdtrack
British Trees
Google Maps (you can securely share your live location with friends/family for safe, happy travels)
iNaturalist (if like me, you have no idea what you're looking at but would like to)
iRecord (see above)
Mammal Mapper
Nature Finder
SeagrassSpotter

Index